Hands-On Visual Studio 2022

A developer's guide to new features and best practices with .NET 8 and VS 2022 for maximum productivity

Hector Uriel Perez Rojas

Miguel Angel Teheran Garcia

Hands-On Visual Studio 2022

Group Product Manager: Rohit Rajkumar

Publishing Product Manager: Vaideeshwari Muralikrishnan

Book Project Manager: Sonam Pandey

Senior Editor: Anuradha Joglekar

Technical Editor: Simran Haresh Udasi

Copy Editor: Safis Editing

Indexer: Rekha Nair

Production Designer: Aparna Bhagat

DevRel Marketing Coordinators: Anamika Singh and Nivedita Pandey

First edition: June 2022
Second edition: March 2024
Production reference: 1310124

Published by Packt Publishing Ltd.
Grosvenor House
11 St Paul's Square
Birmingham
B3 1RB, UK

ISBN 978-1-83508-044-3

www.packtpub.com

To God, first and foremost, for his blessings. To my wife, Cristina, for her patience and love, to my beloved children, Elizabeth and Ricardo, for their antics, which make me laugh every day, and to my parents and siblings, for their unconditional support.

– Hector Perez

To my mother, Micaela Garcia, and my sister, Maria Angelica Teheran, for teaching me the value of hard work, and a special mention to my wife, Rina Plata, for supporting me during the evolution of my professional career and personal life.

– Miguel Garcia

Contributors

About the authors

Hector Uriel Perez Rojas is an experienced senior developer, with more than 10 years of experience in developing desktop, web, and mobile solutions with the .NET platform. He was recognized in 2021 with the Microsoft MVP award. He is an active member of the community and has his own training academy: Devs School.

I want to thank my family for their unconditional support and love, the developer community for always being so active, those who have taught me something valuable throughout my life, and all the people who have made the publication of this book possible.

Miguel Angel Teheran Garcia is a solutions architect expert in Microsoft technologies, with over 10 years of experience. In 2020, he was recognized as a Microsoft MVP and, in 2021, as an Alibaba Cloud MVP. Miguel is an active member of tech communities and a content author with C# Corner.

I want to thank my family, my parents, and friends, who have helped and supported me in every new challenge in my personal life and professional career. And I want to thank people who follow my content and help me to improve it every day.

About the reviewer

Luis Beltrán is an engineer with expertise in the cloud, mobile, artificial intelligence, and software development who has been involved in creating, enhancing, and integrating enterprise solutions throughout his career. Currently, he works as a university lecturer and consultant, leveraging his technical experience to provide valuable insights and solutions for various real-world challenges as well as to shape students' career paths. As part of his ongoing professional growth, he holds several role-based Azure certifications. Microsoft has honored him with the Most Valuable Professional award for 7 years in a row in acknowledgment of his technical proficiency, leadership qualities, online impact, and dedication to tech communities.

Table of Contents

3

Debugging and Profiling Your Apps   63

Part 2: Tools and Productivity

4

Adding Code Snippets   99

5

Coding Efficiently with AI and Code Views   115

6

Using Tools for Frontend and Backend Development   139

7

Styling and Cleanup Tools   165

Part 3: GitHub Integration and Extensions

Preface

Visual Studio 2022 is the best **integrated development environment (IDE)** for .NET developers. The powerful tools added release after release make it an annually renewed IDE for application development across various domains. This book is a comprehensive guide for .NET developers, both beginners and advanced, who wish to familiarize themselves with the basic, advanced, and new features of the IDE. The book is divided into three sections: an overview of Visual Studio, tools for productivity within the IDE, and finally, integration with GitHub and Visual Studio extensions.

In this second edition, we expand the use of Visual Studio to environments beyond web development, such as Azure, Desktop, .NET MAUI, and others. Moreover, updates to the IDE introduced since the release year of the first edition of the book in 2022 have been included, such as the use of GPT-based extensions such as GitHub Copilot and GitHub Copilot Chat, as well as profiling tools.

The book reviews breakpoint groups, a new way to configure breakpoints, as well as tools for web development, such as Browser Link, Dev Tunnels, and Web API Endpoints Explorer. For more advanced developers, the book addresses the topic of developing extensions through the use of the new Visual Studio SDK.

Who this book is for

This book is intended for .NET developers who want to learn how to use the latest features, tools, and extensions available in Visual Studio 2022. To get the most out of the book, it is recommended to have general knowledge of C#. Knowledge of web development, Azure, and .NET MAUI and a desktop with .NET, although not required, would be helpful.

What this book covers

Chapter 1, Getting Started with Visual Studio 2022, reviews how to install Visual Studio, new features, customization, and accessibility options.

Chapter 2, Creating Projects and Templates, explains the templates included in Visual Studio and how to create your first projects using them.

Chapter 3, Debugging and Profiling Your Apps, discusses how to use Visual Studio to build applications, debug your projects, and profile your apps.

Chapter 4, Adding Code Snippets, explains how to use code snippets and create your own in Visual Studio.

Chapter 5, Coding Efficiently with AI and Code Views, reviews the AI tools and different code views that we can use in Visual Studio.

Chapter 6, Using Tools for Frontend and Backend Development, explains some of the tools included in Visual Studio for frontend and backend development.

Chapter 7, Styling and Cleanup Tools, discusses the code cleanup options included in Visual Studio.

Chapter 8, Publishing Projects, explores the different ways to publish different project types from Visual Studio.

Chapter 9, Implementing Git Integration, reviews the Visual Studio functionalities to work with GitHub-hosted projects.

Chapter 10, Sharing Code with Live Share, discusses what Live Share is and how to use it to work with teams in live coding sessions.

Chapter 11, Working with Extensions in Visual Studio, explains what extensions are in Visual Studio, the different ways to add them to the IDE, and how to create your own extensions.

Chapter 12, Using Popular Extensions, discusses what the most popular extensions for Visual Studio are and why.

Chapter 13, Learning Keyboard Shortcuts, explains how to perform quick actions in Visual Studio using the keyboard to optimize repetitive tasks.

To get the most out of this book

You need to have a laptop or desktop computer with Windows 10 or later installed. To install Visual Studio and complete the exercises, you will need an internet connection.

To obtain a Visual Studio Community license, you must have a Microsoft account, either belonging to the Hotmail domain or the Outlook domain.

Software/hardware covered in the book	Operating system requirements
ARM64 or x64 processor; quad-core or better recommended. ARM 32 processors are not supported. Minimum of 4 GB of RAM. Many factors impact the resources used; we recommend 16 GB RAM for typical professional solutions. Hard disk space: Minimum of 850 MB up to 210 GB of available space, depending on features installed; typical installations require 20-50 GB of free space. We recommend installing Windows and Visual Studio on a solid-state drive (SSD) to increase performance.	Windows 10 or later
Visual Studio 2022 Community Edition	Windows 10 or later

A version of Visual Studio Enterprise is required to generate code maps in Chapter 5, Coding Efficiently with AI and Code Views.

A GitHub Copilot license is required to perform the AI completion tests in Chapter 5, Coding Efficiently with AI and Code Views.

To perform the tests suggested in Chapter 9, Implementing Git Integration, a GitHub account is required.

If you are using the digital version of this book, we advise you to type the code yourself or access the code from the book's GitHub repository (a link is available in the next section). Doing so will help you avoid any potential errors related to the copying and pasting of code.

To complete and understand all the activities throughout this book, it's important to have basic knowledge about software development with .NET.

Download the example code files

You can download the example code files for this book from GitHub at `https://github.com/PacktPublishing/Hands-On-Visual-Studio-2022-Second-Edition`. If there's an update to the code, it will be updated in the GitHub repository.

We also have other code bundles from our rich catalog of books and videos available at `https://github.com/PacktPublishing/`. Check them out!

Conventions used

There are a number of text conventions used throughout this book.

`Code in text`: Indicates code words in text, database table names, folder names, filenames, file extensions, pathnames, dummy URLs, user input, and Twitter handles. Here is an example: "In the `custom.css` file, you can write the `columns` property and see how VS suggests a code snippet for it."

A block of code is set as follows:

```
if (OperatingSystem.IsLinux())
{
    return new List<WeatherForecast>();
```

When we wish to draw your attention to a particular part of a code block, the relevant lines or items are set in bold:

```
public float Calculate(int value)
{
    return (float)((value + 126) * (Math.PI / value));
}
```

Bold: Indicates a new term, an important word, or words that you see onscreen. For instance, words in menus or dialog boxes appear in **bold**. Here is an example: "Go to the **Tools | Options** menu."

> **Tips or important notes**
> Appear like this.

Get in touch

Feedback from our readers is always welcome.

General feedback: If you have questions about any aspect of this book, email us at customercare@packtpub.com and mention the book title in the subject of your message.

Errata: Although we have taken every care to ensure the accuracy of our content, mistakes do happen. If you have found a mistake in this book, we would be grateful if you would report this to us. Please visit www.packtpub.com/support/errata and fill in the form.

Piracy: If you come across any illegal copies of our works in any form on the internet, we would be grateful if you would provide us with the location address or website name. Please contact us at copyright@packt.com with a link to the material.

If you are interested in becoming an author: If there is a topic that you have expertise in and you are interested in either writing or contributing to a book, please visit authors.packtpub.com.

Share Your Thoughts

Once you've read *Hands-On Visual Studio 2022*, we'd love to hear your thoughts! Scan the QR code below to go straight to the Amazon review page for this book and share your feedback.

https://packt.link/r/1835080448

Your review is important to us and the tech community and will help us make sure we're delivering excellent quality content.

Download a free PDF copy of this book

Thanks for purchasing this book!

Do you like to read on the go but are unable to carry your print books everywhere?

Is your eBook purchase not compatible with the device of your choice?

Don't worry, now with every Packt book you get a DRM-free PDF version of that book at no cost.

Read anywhere, any place, on any device. Search, copy, and paste code from your favorite technical books directly into your application.

The perks don't stop there, you can get exclusive access to discounts, newsletters, and great free content in your inbox daily

Follow these simple steps to get the benefits:

1. Scan the QR code or visit the link below

https://packt.link/free-ebook/9781835080443

2. Submit your proof of purchase
3. That's it! We'll send your free PDF and other benefits to your email directly

Part 1: Visual Studio Overview

In this part, you will learn how to install and use Visual Studio 2022 from scratch and about the general tools provided by this integrated development environment.

This part has the following chapters:

- *Chapter 1, Getting Started with Visual Studio 2022*
- *Chapter 2, Creating Projects and Templates*
- *Chapter 3, Debugging and Profiling Your Apps*

1

Getting Started with Visual Studio 2022

Visual Studio (VS) is the most popular **integrated development environment (IDE)** for **.NET** developers. It's the perfect tool for designing, developing, debugging, and deploying all .NET applications and even other technologies.

This book has been written concerning the latest developments and features added to the IDE, which are very interesting and useful to developers of all platforms. Throughout this book, you will discover the different tools and options of the IDE to develop applications in general, thus helping you make it your own. This will make become your most powerful ally when creating applications.

In this chapter, you will learn about the evolution of VS as well as the improvements introduced in the new version. After, you will learn how to install the IDE, take a general tour of the graphical interface, customize it according to your preferences, and explore the accessibility features included in the IDE.

By the end of this chapter, you will have a general understanding of the VS ecosystem, including the new features that have been added to the latest version. You will also know how to install and customize the IDE according to your needs, including selecting themes and adjusting colors and fonts for the code editor. Additionally, you will know where to find tools in the graphical interface and what accessibility options are included in the environment.

In this chapter, we will cover the following main topics:

- A brief history of VS and its flavors
- Improvements in VS 2022
- Installing VS 2022
- Introduction to the IDE
- Customizing the IDE
- Accessibility features

Technical requirements

We will begin this chapter by learning how to install VS 2022. To get VS to run on your machine, you will need the following:

- Windows 10 minimum supported operating system version or higher: Home, Professional, Education, or Enterprise

- Windows 11 minimum supported operating system version or higher: Home, Pro, Pro Education, Pro for Workstations, Enterprise, or Education

- Windows Server, 2016 or higher

- A 1.8 GHz or faster 64-bit processor; quad-core or better is recommended

- 4 GB of RAM; 8 GB is recommended

- Hard disk free space – 25 GB (up to 40 GB depending on the components installed)

- Administrator rights

- Full internet access during the installation

> **Important note**
>
> 32-bit and ARM operating systems are not supported; you will need either Windows 10 Enterprise LTSC edition, Windows 10 S, or Windows 10 Team Edition. To check all the requirements and technologies or systems not supported, visit `https://docs.microsoft.com/en-us/visualstudio/releases/2022/system-requirements`.

A brief history of VS and its flavors

VS 2022 is version 13 of this application created by Microsoft. VS has been consolidated among developers for having a friendly user experience, good support with regular updates, and powerful tools for writing clean and scalable code. VS has support for many technologies and platforms. For many developers, VS is the ultimate tool for the creation of web, mobile, and desktop projects, as we will see throughout this book.

To understand the evolution of VS, we must examine its history and timeline.

VS 6.0 was released in 1997, and it was the first version of this tool. This version was created to work with Visual Basic 6.0. Then, in 2002, a new version was released, which included compatibility with **.NET** and **C#** (a new programming language at that time). Since then, it's been the favorite tool for .NET developers.

Since the 2012 version, the development team has implemented a new look and feel and introduced many improvements in the user experience, which are also present in the 2022 version. Some of the most important improvements in VS 2012 over the previous versions were performance, the possibility to choose from light and dark themes, and a new set of icons that have replaced those of previous versions.

Another turning point of the 2012 version was the introduction of three flavors that cover all developers' preferences and needs:

- VS Community
- VS Professional
- VS Enterprise

However, only VS Community is completely free for the community. So, let's understand the main aspects of each version.

VS Community

VS Community is a free version that incorporates all the basic tools to create, build, debug, and deploy .NET applications and all the collaboration instruments integrated into VS.

VS Community has a limit of five users and is restricted to non-enterprise organizations. This version is suitable for students, independent developers, freelancers, and small companies.

The main tools in VS Community are as follows:

- **Basic debugging tools**: Tools for inspecting code during debugging
- **A performance and diagnostics hub**: Tools for analyzing application performance and memory use
- **Refactoring tools**: Tools for cleaning and styling code following best practices
- **Unit testing**: A feature to navigate, run, and collect results from unit tests
- **Peek definition**: A functionality to navigate to the definition of a method or function
- **VS Live Share**: A tool for real-time collaboration development

Even though this version includes all the main tools that you will use daily, in some scenarios associated with unit testing, memory, or inspection, these tools aren't enough. For example, in VS Community, it is not possible to use Live Unit Testing, a feature that allows you to get feedback from unit tests in real time, as they are constantly running in the background in VS.

VS Professional

VS Professional is a licensed version of VS offered by subscription; this version is recommended for enterprise applications and teams with more than five developers. VS Professional includes the same tools as VS Community but with some additions, such as **CodeLens** (a VS feature to find references, changes, and unit testing in code).

At the time of writing, the cost of the VS Professional subscription is $45 per month for an individual user.

The Professional subscription includes $50 in Azure credits monthly, training, support, and Azure DevOps (basic plan).

VS Enterprise

VS Enterprise is the top-level subscription version of VS (with VS Professional) that includes all of VS Community's features, VS Professional's improvements, and some additional tools, such as **IntelliTrace**, which records the state of the application to be able to perform a debugging by inspecting the system before an exception in detail, knowing information about events, exceptions, and function calls.

Some features to highlight are as follows:

- **Live unit testing**: To rerun unit testing automatically after making a change
- **Snapshot Debugger**: A tool for saving snapshots during debugging when an error occurs
- **Performance analysis tools**: For mobile applications
- **Architectural layer diagrams**: To visualize the logical architecture of your app

The Enterprise subscription has a $250 fee per month, but it includes $150 in Azure credits monthly, Power BI Pro, Azure DevOps with test plans, and all the features available for VS.

To see a comparison of the different flavors and prices, you can go to `https://visualstudio.microsoft.com/vs/pricing/`.

> **Important note**
>
> For this book, we are going to use **VS Community**. Since this is a free version, you don't have to pay any subscription, and all the topics are covered with this version.

Now that you've been provided with a brief history of the IDE and the versions available, let's look at some of the new enhancements in VS 2022.

Improvements in VS 2022

Significant enhancements have been introduced in VS 2022. One of the most important is the transition to a 64-bit architecture, which introduces a significant performance improvement.

Let's take a closer look at the specific enhancements.

A 64-bit architecture

A simple but important feature in VS 2022 is the new architecture, which is 64-bit. This is a change that we cannot see, but internally, it takes advantage of a 64-bit CPU, which is common in current laptops and PCs, to improve performance and reduce delays in the execution of multiple tasks.

Using **Task Manager** in Windows, you will be able to notice the difference when VS 2019 and 2022 are running at the same time:

Name	PID	Status	User name	CPU	Memory (a...	Archite...	Description
AggregatorHost.exe	2912	Running	SYSTEM	00	144 K	x64	Microsoft (R) Aggregator Host
conhost.exe	2440	Running	SYSTEM	00	60 K	x64	Console Window Host
csrss.exe	632	Running	SYSTEM	00	568 K	x64	Client Server Runtime Process
csrss.exe	704	Running	SYSTEM	00	156 K	x64	Client Server Runtime Process
csrss.exe	1872	Running	SYSTEM	00	748 K	x64	Client Server Runtime Process
ctfmon.exe	7840	Running	hprez21	00	2,308 K	x64	CTF Loader
devenv.exe	6156	Running	hprez21	05	397,592 K	x64	Microsoft Visual Studio 2022
devenv.exe	7128	Running	hprez21	17	217,176 K	x86	Microsoft Visual Studio 2019
dllhost.exe	4128	Running	hprez21	00	2,028 K	x64	COM Surrogate
dllhost.exe	6684	Running	hprez21	00	604 K	x64	COM Surrogate
dwm.exe	536	Running	DWM-1	00	3,684 K	x64	Desktop Window Manager
dwm.exe	4320	Running	DWM-2	00	21,908 K	x64	Desktop Window Manager
explorer.exe	1880	Running	hprez21	00	21,848 K	x64	Windows Explorer
fontdrvhost.exe	956	Running	UMFD-1	00	68 K	x64	Usermode Font Driver Host
fontdrvhost.exe	964	Running	UMFD-0	00	180 K	x64	Usermode Font Driver Host
fontdrvhost.exe	3988	Running	UMFD-2	00	584 K	x64	Usermode Font Driver Host
LocationNotification...	5996	Running	hprez21	00	44 K	x64	Location Notification
LogonUI.exe	4592	Running	SYSTEM	00	4,332 K	x64	LogonUI
lsass.exe	844	Running	SYSTEM	00	3,432 K	x64	Local Security Authority Process
Microsoft.ServiceHu...	4452	Running	hprez21	00	21,776 K	x64	Microsoft.ServiceHub.Controller
Microsoft.ServiceHu...	9428	Running	hprez21	00	18,632 K	x64	Microsoft.ServiceHub.Controller
msedge.exe	5736	Running	hprez21	00	31,192 K	x64	Microsoft Edge
msedge.exe	5404	Running	hprez21	00	272 K	x64	Microsoft Edge
msedge.exe	7316	Running	hprez21	00	10,940 K	x64	Microsoft Edge

Figure 1.1 – VS 2022 on the 64-bit platform and VS 2019 on the 32-bit platform

With a 32-bit architecture, there was an access limitation of 4 GB of memory. Now, thanks to a 64-bit architecture, it is possible to access a larger amount of memory, whose limit depends on the memory in the physical equipment, reducing time limits and avoiding IDE freezes or crashes.

All in all, the VS development team improved the performance for many scenarios in the 2022 version. You will notice the difference when working on large projects, and performance will be better in future versions.

A 64-bit architecture is a good improvement for performance, but this feature doesn't improve user interaction while coding. In the following section, we will see how the icons and style were improved to have a better user experience.

New icons and styles

New icons and styles were added in VS 2022. Although this is a simple feature, it helps us to navigate easily in an application, using visual elements and identifying actions and tools properly. For instance, in *Figure 1.2*, you can see that the broom icon (row one, column five) on the right has better contrast, with a new vibrant yellow color (specifically for a dark theme) and a modern design:

Figure 1.2 – Icons in VS 2022 versus VS 2019

The broom icon is used to execute code cleanup to fix code formatting. The current icon also implies that code cleanup will be executed.

> **Important note**
>
> An interesting fact about the new icons in VS 2022 is that the VS development team worked with the developer community to fulfill three purposes – consistency, readability, and familiarity. This resulted in a series of icons with the same meaning but with consistent colors, sharp contrast, and a recognizable silhouette.

The contrast between letters, icons, and the background was improved to make it more pleasant and less tiring for the eyes. *Figure 1.3* shows an example of this:

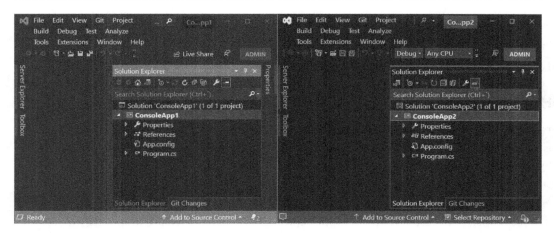

Figure 1.3 – Example of the dark theme in VS 2019 (left) versus VS 2022 (right)

In *Figure 1.3*, we can see a contrast between VS 2019 and VS 2022, with the new version being a better combination of minimalist icons and providing a better visual appearance.

> **Important note**
>
> While the use of dark themes is becoming more common among developers in general, it is also becoming more prevalent in applications across the industry.

There were also some changes in the other themes, but dark and blue were the most improved.

With these new icons and styles, working with VS becomes more user-friendly. In the next section, we will see how VS supports the new version of .NET, known as .NET 8.

.NET 8 support

.NET is a free, open source framework that's used to create web, desktop, mobile, and other kinds of applications using C#, F#, and Visual Basic (with C# being the most popular).

A new version of **.NET** is released every year. .NET 8 is a new long-term support version that offers great new improvements, such as Native AOT for apps and C# 12 compatibility. For more information about these improvements, visit `https://docs.microsoft.com/en-us/dotnet/core/whats-new/dotnet-8`.

VS 2022 is ready to create, compile, and publish projects using this new version.

In *Chapter 2, Creating Projects and Templates*, we will analyze some projects and templates provided by VS and we will look at the option to choose .NET 8 for our applications.

We can create projects using .NET 8. Additionally, we can use a new functionality for checking the code changes known as **Hot Reload**.

Let's review how Hot Reload can improve our productivity using VS 2022.

Hot Reload

For a long time, **Hot Reload** was the main requested feature in the .NET developers' community. This is a feature that's already implemented in many technologies and expands a developer's productivity by refreshing an application after every change made to the code.

> **Important note**
> Hot Reload is a feature that recompiles code after a change is made to it. This way, an application displays visual changes immediately without the need to restart it, significantly increasing productivity.

VS 2022 includes this amazing feature for many kinds of projects, including ones involving ASP.NET, .NET MAUI, Windows Forms, and WPF:

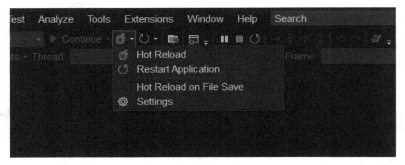

Figure 1.4 – The Hot Reload button in VS 2022

In *Chapter 6, Using Tools for Frontend and Backend Development*, we will use this utility in some demos and learn why this tool is very important to us. In the next section, we will share other improvements that can help us increase our productivity, especially in larger projects.

Other improvements

VS 2022 has other improvements that you will discover as the chapters go by since learning how to use them sometimes requires more space than a couple of paragraphs.

Among the most outstanding are a big improvement in the **Search File** tool to find elements quickly in a project with many files, the addition of several types of breakpoints for better debugging, the use of GPT-based tools to interact with our project through artificial intelligence, the addition of web features for web API testing, and the introduction of new tools to work with Git repositories, among many others.

Now that you know about some of the improvements that were made in VS 2022, let's learn how to install the IDE on our machine.

Installing VS 2022

The VS installation process has improved with each new version that has been released. Today, it is even possible to reuse the same installer to perform upgrades or make workload modifications to create different types of projects.

Getting the installer from the website

VS is easy to install, and you don't need an account or take a lot of steps to get the installer. Simply go to the following link, where you can get the installer directly: `https://visualstudio.microsoft.com/downloads/`.

On the web page, go to the **Visual Studio** section, click on the drop-down control marked **Download Visual Studio**, and select the **Community** option (see *Figure 1.5*):

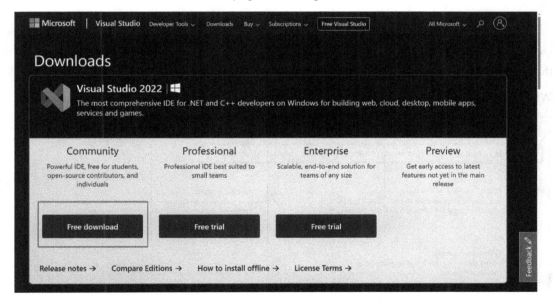

Figure 1.5 – Downloading VS

> **Note**
>
> If the download site shows a new version of VS, you can find the download links to VS 2022 and other previous versions at `https://visualstudio.microsoft.com/vs/older-downloads/`.

Once you have downloaded the VS installer, you must run it to start the update process of the installer itself, as shown in *Figure 1.6*:

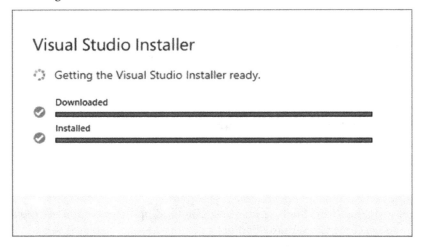

Figure 1.6 – Updating the VS installer

VS is an IDE that is constantly updated by the development team, so the installer will always look for the latest available update to perform the corresponding installation.

When the installer update has been completed, the initial screen of the installer will be presented. In the top menu, you will see four options to configure the IDE, as shown in *Figure 1.7*:

Figure 1.7 – The VS Workloads tab

Here is a brief description of each tab in the installer:

- **Workloads**: From this tab, it is possible to find workloads for different types of projects – for example, mobile projects, desktop-focused projects, and Python-focused projects, among others. Each workload includes a set of tools and components associated with the technology selected.

- **Individual components**: This tab shows the list of individual components that will be installed according to the workload. We can select components individually, even if they do not belong to the selected workloads.

- **Language packs**: Here, you will be able to select the language or languages of the VS interface from a list of 14 languages. This is very useful since the default language usually corresponds to the language in which the VS installer has been downloaded. From here, you will be able to deselect the default language and select a different one or multiple languages to switch between in your development process.

- **Installation locations**: In this tab, you will be able to see how much space is needed in each of the paths to perform a correct installation.

Finally, there is a list of options to choose the installation method – whether you want to proceed to download all the components and install them at the same time, or you want to download all the necessary components first and install them later (*Figure 1.8*):

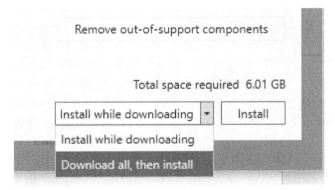

Figure 1.8 – The VS Install button

If you agree with the details and size of the installation, you can start the process by clicking on the **Install** button.

> **Note**
> It is possible to modify workloads, individual components, and user interface languages by re-running the same installer that has been downloaded.

Once you have clicked the button to start the installation, a window will appear showing the details of the download and installation of the components, as shown in *Figure 1.9*:

Figure 1.9 – The VS installation in progress

Now that you know how to install VS on your computer, let's take a brief tour of the interface.

Introduction to the IDE

Now that VS 2022 is installed, you will be shown the startup screen for creating, cloning, and opening projects. From here, you can check that the installation has been successful, as shown in *Figure 1.10*:

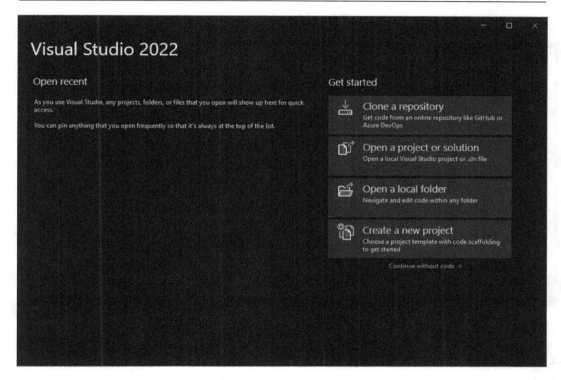

Figure 1.10 – The VS startup window

> **Important note**
>
> You can have different versions of VS on the same machine; for compatibility with all technology, sometimes, we need to keep old versions – for example, VS 2010 to work with Silverlight (an unsupported framework that's used to create web applications with C# and XAML, which was executed through a plugin in the browser).

Let's go directly to the interface without creating any project yet via the **Continue without code** option. Once you are in the main interface of VS, at the top, you will be able to see the **Options** menu, which provides a wide range of tools for different purposes, as well as buttons to perform common actions quickly, such as starting the execution of an application, debugging tools, and more, as shown in *Figure 1.11*:

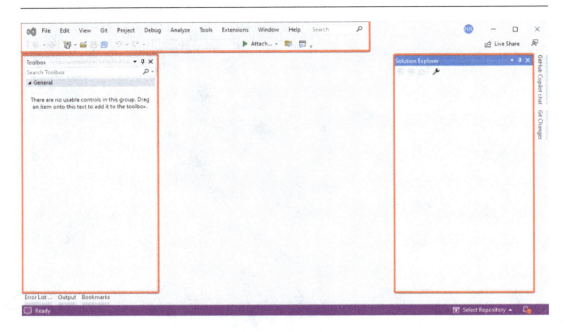

Figure 1.11 – VS general interface

It should be noted that the tools that appear may change according to the open project. In this case, for instance, since we have not created any project yet, the toolbar appears with a minimal amount of options.

On the other hand, in the left and right sections, we can see something known as panels. We are going to analyze these in the next subsection.

Customizing panels

Panels in VS are a way to access specific tools according to the type of project you are working on.

These panels are composed of tools and document editors, some of which are used most of the time, such as **Solution Explorer** (to see the structure of your projects), **Toolbox** (which shows you controls to drag and drop according to the current project), **Properties** (to modify the properties of the selected element), and **Code Editor**.

Let's learn how to configure the IDE so that you can see which tools you will need for your project.

Adding tools to panels

VS 2022 has many tools or windows that you can show or hide according to your needs. This list of tools can be found in the **View** menu, as demonstrated in *Figure 1.12*:

⊡	Solution Explorer	Ctrl+W, S
⬚	Git Changes	Ctrl+0, G
⬚	Git Repository	Ctrl+0, Ctrl+R
⬚	Team Explorer	Ctrl+|, Ctrl+M
⊟	Server Explorer	Ctrl+W, L
⬚	SQL Server Object Explorer	Ctrl+|, Ctrl+S
⬚	Test Explorer	
⬚	GitHub Copilot chat	Ctrl+W, I
⬚	Bookmark Window	Ctrl+W, B
⬚	Call Hierarchy	Ctrl+W, K
⬚	Class View	Ctrl+W, C
⬚	Code Definition Window	Ctrl+W, D
⬚	Object Browser	Ctrl+W, J
⬚	Error List	Ctrl+W, E
⬚	Output	Ctrl+W, O
⬚	Task List	Ctrl+W, T
⬚	Toolbox	Ctrl+W, X
⬚	Notifications	Ctrl+|, Ctrl+N
⬚	Terminal	
	Other Windows	▸
	Toolbars	▸

Figure 1.12 – The list of available tools

Once you have opened this menu, you will find the tools that are usually the most used listed immediately. These include **Server Explorer**, **Class Viewer**, **Error Listing**, **Output Window**, and **Terminal**, among others.

There is another set of tools that are not as widely used but may help you at some point. These can be found in the **Other Windows** section.

From here, you can access tools such as **Containers, C# Interactive, Data Sources**, and **Package Manager Console**, among others.

To add any of these tools to one of the panels, simply select one of them; it will automatically be added to your current environment in a strategic panel. For example, if you add the **Server Explorer** tool, it will be added to the left panel. On the other hand, if you add the **Output** tool, it will be added to the bottom pane.

Panel accommodation

A great advantage of VS is that you can arrange the tool panels wherever you prefer. To achieve the best results, it is convenient that you know the structure of a panel.

Each panel is composed of five sections where you can place tools. These sections are located on each side of the panel, plus one in the center, as shown in *Figure 1.13*:

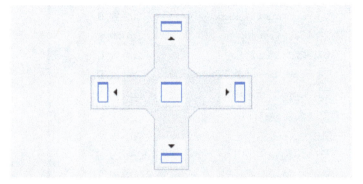

Figure 1.13 – The sections of a panel

To place a tool in a panel, simply position the cursor at the top of the tool and slide it to the panel you want. The IDE itself will show you the possible locations where you can place the tool, making the process simple and easy; you can also pull the panel out of the main windows and use it on its own. *Figure 1.14* shows what the process of docking a window looks like:

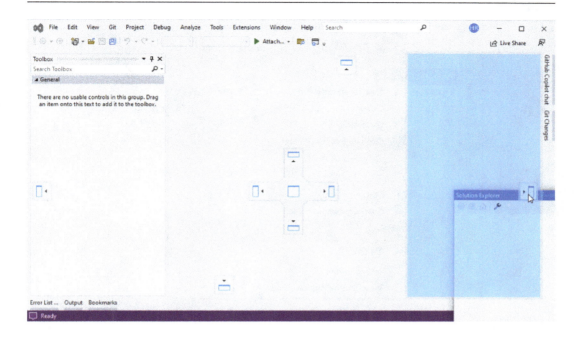

Figure 1.14 – Docking a window in a panel

Now, it's time to review how we can work with documents.

Working with documents

There are special options that we can apply when working with document editors, such as **Code Editor**. If we want to see these options, we just need to right-click on the tab of the open document, as shown in *Figure 1.15*:

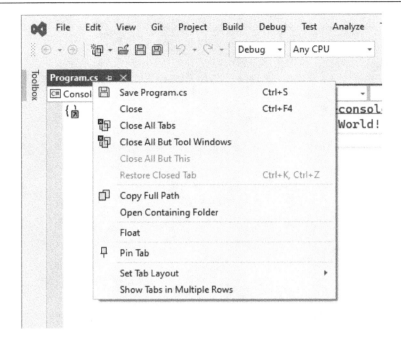

Figure 1.15 – The context menu for Code Editor

These options are quite intuitive – for example, the **Float** option will allow us to turn the editor into a floating window, which we can drag to a second monitor. The **Pin Tab** option will allow us to set the tab at the beginning of the open windows, while the **Set Tab Layout** option will allow us to move the set of tabs to the left, the top, or the right.

It is important to highlight that if we have more documents open, we will have additional options available. With this option, we can create groups of documents to distribute the space and use it to perform tasks to increase productivity, such as comparing two documents. *Figure 1.16* demonstrates this visually:

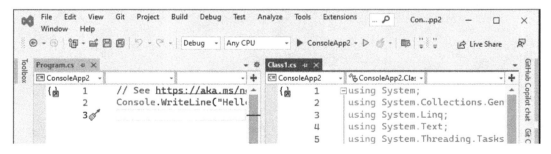

Figure 1.16 – Document groups

> **Tip – a help shortcut**
> Use *Ctrl + F1* to open the VS documentation to read guides and news.

Now that you know about the main elements of the graphical interface, it is time to learn how to customize the IDE.

Customizing the IDE

In general, it is common for software developers to have different tastes in writing code. This also applies in the configuration part of the tool you use to develop, such as whether you want to change the general color of the IDE or prefer to change the default font to one that suits you better. In this section, we'll look at some of the customizations available in VS 2022.

Selecting a VS theme

The first time you start VS 2022, you will be presented with a dark window, as shown in *Figure 1.17*. This is because the default dark theme has been applied:

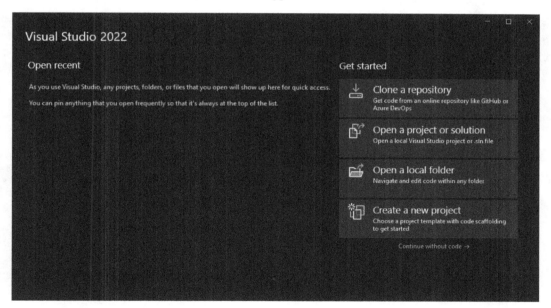

Figure 1.17 – VS default theme

Although the dark theme has been applied by default, there are several pre-installed themes, as shown in *Figure 1.19*:

- **Blue**
- **Blue (Extra Contrast)**
- **Dark**
- **Light**
- **Use system setting**

Each of these options has a set of pre-established colors that you will be able to preview using the same window.

> **Important note**
>
> The **Dark** theme helps reduce eye strain in low-light conditions. This is a perfect option if you need to work for many hours per day in an office or places with limited light. *Some figures and screenshots will be in light mode in this book.*

If you want to change the theme you selected at the beginning, you can do so by going to the main VS window and clicking on the link that says **Continue without code**:

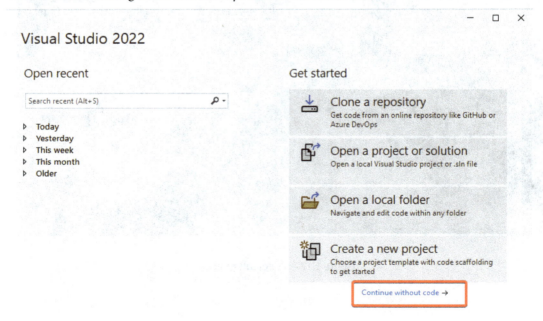

Figure 1.18 – Accessing VS without code

Then, from the **Tools** drop-down menu, you can go to the **Themes** section; you will find all the default themes and those that you have previously installed. You only need to select one to apply the selected theme, as shown in *Figure 1.19*:

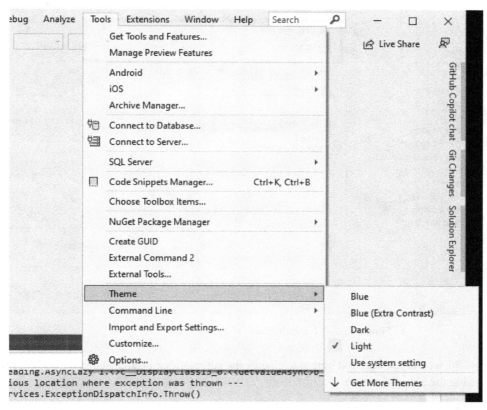

Figure 1.19 – Changing the VS theme

You can also see an additional option called **Get More Themes**. This will take you to the VS marketplace, where you can get themes created by other developers for free.

Now that you know how to choose the theme of your choice, let's learn how to customize fonts and other elements.

Customizing fonts and colors in the environment

To change typography at the IDE level, follow these steps:

1. Go to the **Tools | Options** menu.
2. In the configuration window, go to the **Environment | Fonts and Colors** section.

3. In this section, select the **Environment** option in the **Show settings for** drop-down list.

    This will allow you to change options such as font and size:

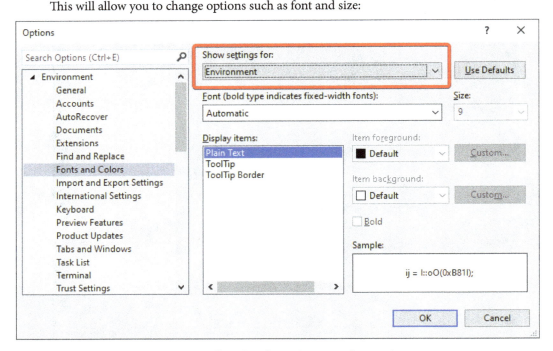

Figure 1.20 – Changing the environment font settings

In *Figure 1.20*, we can use the **Font** option to change the automatic or default font in VS and see how it looks using the **Sample** section.

If you want to change the settings for **Code Editor**, just choose the **Text Editor** option from the drop-down menu. As part of the configuration options, you can change the font used, the font color, the font size, and the background, among other attributes.

One of the advantages of this configuration option is that it allows you to be as specific with the configuration as you need. This means that you can also alter the typography for things such as selected text, line numbers, bookmarks, and code snippets, among many other settings.

> **Important note**
> In VS 2022, the Cascadia font was introduced as the default font. It has been described as a fun font with better code readability.

Now, let's see how VS can help us maintain our preferences across different computers thanks to account synchronization.

Synchronizing accounts and settings

A great feature of VS is that it allows you to synchronize the configurations you make and allows you to comfortably work on different computers.

This is possible thanks to a Microsoft account, which is required for the use of VS. This account is requested when you start VS for the first time, though you can enter the account or modify it for use in VS at any time by clicking on the **Sign in** option, located at the top right of the IDE, as shown in *Figure 1.21*:

Figure 1.21 – VS – Sign in (at the top right of the main screen)

The main configurations that are synchronized through this process are as follows:

- User-defined window layout configurations
- Themes and menu settings
- Fonts and colors
- Keyboard shortcuts
- Text editor settings

If you do not want to apply the synchronization of your configuration on a particular computer, it is possible to do so by going to the **Tools | Options | Environment | Accounts** menu. From here, you will be able to deselect the **Synchronize Visual Studio settings across devices** option, as shown in *Figure 1.22*:

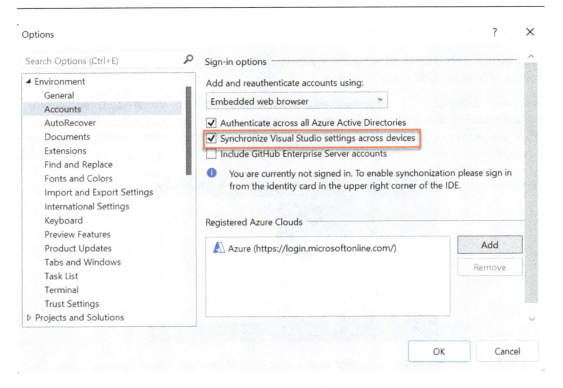

Figure 1.22 – Disabling the synchronization of preferences between devices

> **Important note**
> Disabling this feature will not affect the synchronization process of other versions or editions of VS that may be installed on the same computer.

Now that you've learned how to synchronize the settings made between different computers, let's learn how to customize the menu bar and toolbars.

Customizing the menu bar and toolbars

Menus and toolbars are excellent ways to access tools or options, best known as commands, that you use frequently, so it is very important to learn how to customize them so that they can aid you with developing your applications. In this section, we will cover the following topics:

- Customizing the menu bar
- Customizing toolbars

First, let's learn how to customize the menu bar.

Customizing the menu bar

The menu bar is the set of options that appear at the top of the IDE. They allow you to access a drop-down menu of options (such as the **File** menu) to execute a certain task, display tools, or apply a change to a project:

Figure 1.23 – The default menu bar in VS 2022

If you want to change the tools that are part of the initial configuration, either to add options in a specific menu or to create your own menus, perform the following steps:

1. Go to **Tools | Customize**.

2. Go to the **Commands** section.

3. In this section, you should work with the **Menu bar** option, which will allow you to modify a main menu bar and a secondary menu bar, which you can differentiate with pipeline symbols ("|") in the dropdown, as seen in *Figure 1.24*:

Figure 1.24 – The Menu bar customization window

4. Once you have selected the menu you wish to modify, a preview will appear, showing you how the menu currently looks. From here, you can perform different actions, such as adding commands and menus, deleting menu items, and modifying the order of items, among other options. For example, you can add new commands to the menu bar by clicking the **Add Command** button.

    This will open a new window that will show you each of the commands grouped by category. You can select these and add them to the selected menu, as shown in *Figure 1.25*:

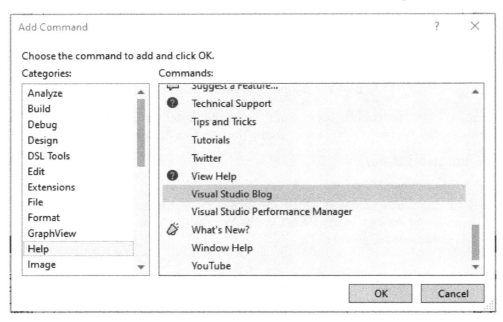

Figure 1.25 – Changing the highlight references options

If you want to remove an option that you've added to the menu, just select it and click on the **Delete** button, as highlighted by a rectangle in *Figure 1.26*:

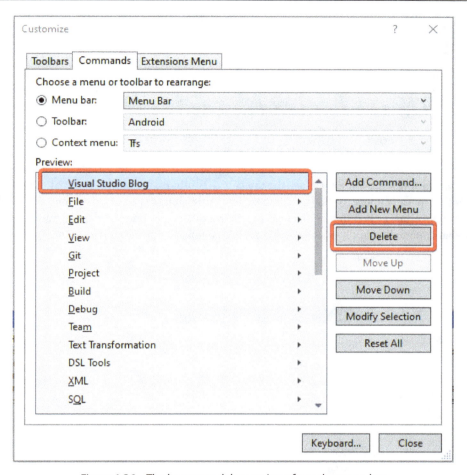

Figure 1.26 – The button to delete an item from the menu bar

> **Important note**
> It is possible to delete toolbars created by the user but not those that are part of the default configuration.

Next, let's see how we can customize the tool bar.

Customizing toolbars

The toolbar is the set of commands that you can access directly without the need to open a menu first, as shown in *Figure 2.17*:

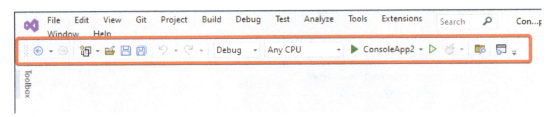

Figure 1.27 – The VS 2022 toolbar section

It is also possible to configure which commands will appear in different tool groups by selecting **Tools | Customize**. In this window, we will see the tab called **Toolbars**, which will show us the different categories that we can choose to show in the interface. By default, the **Standard** option is selected for API and web projects. Depending on the project, other toolbars are added by default, but we can add other toolbars manually, simply by selecting them with a tick, as shown in *Figure 1.28*:

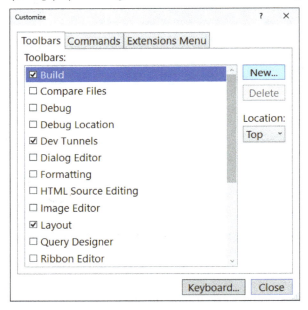

Figure 1.28 – The customization window for adding toolbars to the toolbar section

You can add new commands to a toolbar by going to the **Commands** tab. In this tab, we will select the toolbar we want to modify and carry out the same steps described in the *Customizing the menu bar* section.

If you want to quickly add different toolbars to the IDE, *Figure 1.29* shows how to do it easily by accessing the **View** menu, then **Toolbars**, where you can select and deselect the toolbars you are interested in:

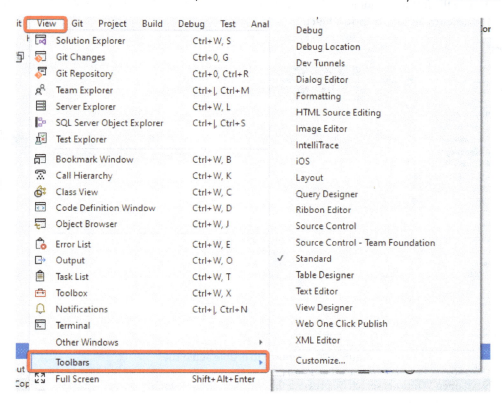

Figure 1.29 – Gaining access to Toolbars from the View menu

Now that you've learned how to customize the IDE to fit your needs, it's time to talk about some accessibility options in VS.

Accessibility features

VS includes several features that make it more friendly to those who are visually impaired.

First, it includes a blue theme with extra contrast, as shown in *Figure 1.19*. This theme helps maximize readability and reduce fatigue for people with vision problems such as color blindness or those with eyestrain and may even be useful for developers who spend a lot of time in front of the computer.

Another feature that has been recently introduced is the ability to add sounds when certain events happen in the IDE. For example, you could add a sound to alert you when the execution of an application reaches a breakpoint, or when there is an error in your code.

To add sounds, follow these steps:

1. Enable the feature by going to **Tools** | **Options** | **Text Editor** | **General**, and enable the **Enable audios cues** option, as shown here:

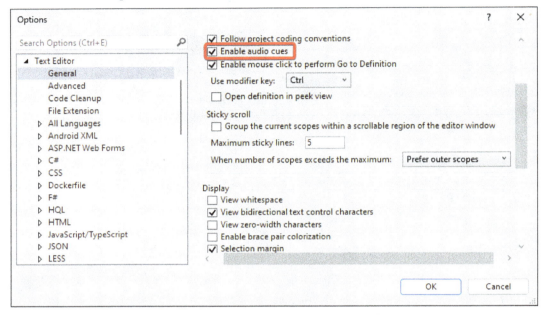

Figure 1.30 – Window to enable audio cues

2. Next, open Windows search by pressing the *Start* key. In the search box, type Change system sounds and open the window, as shown in *Figure 1.31*:

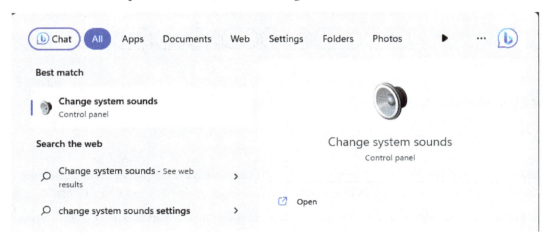

Figure 1.31 – Searching for the tool to change system sounds

3. To finish this process, navigate the tree until you find the **Microsoft Visual Studio** option, where you can assign a sound for the available actions through the **Sounds** drop-down list:

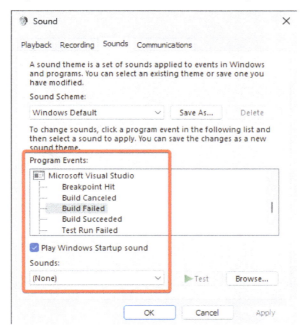

Figure 1.32 – Window for assigning sounds to events in VS

Finally, VS has many shortcuts for accessing common tools through the various keystroke combinations. We'll cover this in detail in *Chapter 13, Learning Keyboard Shortcuts*.

Summary

In this chapter, we provided a brief history of VS, as well as the improvements that were introduced in this new version of VS. Next, you learned how to install VS, as well as how to work with the panels and windows to suit your style and way of working.

You also learned how to customize the IDE, both in terms of the theme and the fonts used in **Code Editor** and the VS **graphical user interface** (**GUI**). You also know how to customize the toolbars and menu bars. Finally, you studied some accessibility features that you can use in case you or a colleague needs them.

In the next chapter, *Chapter 2, Creating Projects and Templates*, you will learn how to create your first project in VS and understand the most used templates and the types of projects you can create.

2

Creating Projects and Templates

A template within VS is a set of files, references, project properties, and compilation options for working with a particular technology. Templates provide us with basic code to work with and act as a guide that we can follow and complete by including our business logic and requirements. Different templates are installed according to the workloads that have been selected during VS installation, although there are templates that will be installed by default, such as class libraries. Depending on the project or technology we want to use, we will find different template options to choose from. Selecting the right template that fits your needs is the best action that you can take to evade technical debts and future issues in your architecture.

In this chapter, we will analyze the main templates provided by VS. In addition, we will understand how to pick the best template for our projects, considering the scope, requirements, and expertise of the team.

We will review the following main topics in this chapter:

- Selecting and searching for templates
- Templates for web development
- Templates for multiplatform development
- Templates for desktop development

Let's see what these templates are all about and how to work with them.

Technical requirements

To follow along in this chapter, you must have previously installed VS 2022, as described in *Chapter 1, Getting Started with Visual Studio 2022*.

You must also have the **ASP.NET and web development**, **.NET Multi-platform App UI development**, and **.NET desktop development** workloads installed. Throughout the chapter, you will see what types of templates are available according to the installed workload.

Selecting and searching for templates

As mentioned in the introduction of this chapter, VS has many templates that we can use with .NET and other technologies, depending on the type of project you are working on.

To explore the templates in VS 2022, just open VS and select the **Create a new project** option:

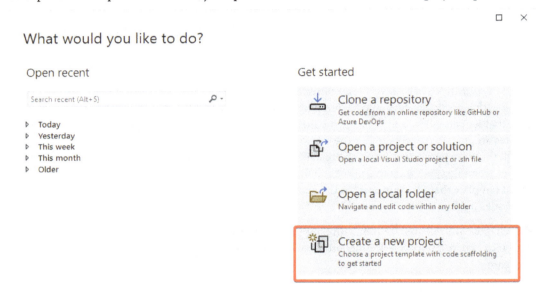

Figure 2.1 – The VS "Create a new project" option

After selecting this option, we will see a list of templates to choose from for creating our new project, as shown in *Figure 2.2*. Similarly, on the left side, we will be able to see templates we have recently selected, with the option to pin or unpin each template so that we can quickly select templates that we use the most:

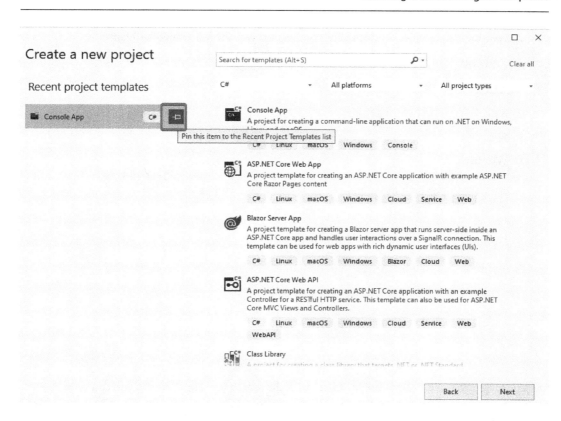

Figure 2.2 – The option to pin or unpin a template in the "Recent project templates" section

In *Figure 2.2*, we also see a search bar at the top for filtering the different templates. If you know the name of the technology that you will use, you can type the term related to it and start searching for the template. The search engine will show you all templates that include the terms you have entered.

Below the search bar, we can also see three drop-down controls, which also serve as filters to search for templates related to the programming language, platforms, or types of projects we want to create. Among the most important options for each of these filters, we can find the following:

- **All languages**: Includes templates to work with the following languages:

  - **C#**
  - **JavaScript**
  - **Python**
  - **TypeScript**
  - **Visual Basic**

- **All platforms**: Includes templates for creating apps on the following platforms:

  - **Android**

  - **Azure**

  - **iOS**

  - **Linux**

  - **Windows**

- **All project types**: Includes templates to filter the following project types:

  - **Cloud**

  - **Desktop**

  - **Games**

  - **Machine Learning**

  - **Mobile**

  - **Test**

  - **Web**

In addition, it is possible to combine the filters with the search bar. For example, you can type `.NET Core` and select the **C#** option in the **All languages** drop-down menu to get all projects related to .NET Core and C#, as shown in *Figure 2.3*:

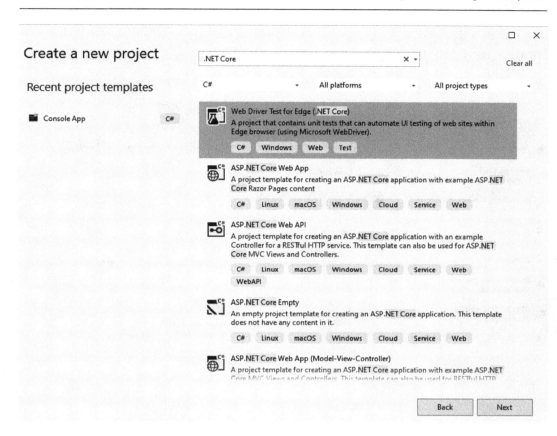

Figure 2.3 – The .NET Core templates for C#

In *Figure 2.3*, we can see many kinds of projects related to .NET Core that VS has found, according to the filters and workloads selected during installation. Additionally, below the description of each template, tags related to the type of project to be created are shown, which are used to fill in information for the search filters.

At the end of the results, you will have an option to install other templates if you cannot find the option you are looking for:

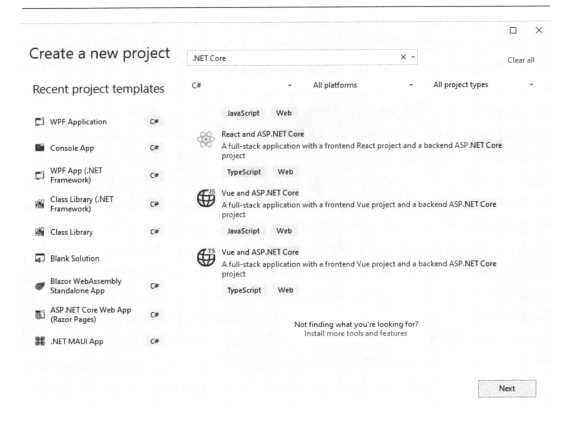

Figure 2.4 – The option to install more tools and features in VS

It is important to note that each template has a description including details related to the template, so we can easily identify whether the template includes the structure and schema that we need.

Another key aspect to consider is that when the description of the template contains the `Empty` keyword, it means that the template doesn't contain elements or modules by default and includes either demos or examples, such as a guide, for the developer. This kind of template only includes the project and the base components to compile and run.

Finally, every time we create a new project from a template, we will be shown a window as depicted in *Figure 2.5*, in which we must specify the following:

- **Project name:** The specific name of the project, not to be confused with the name of the solution. A solution can contain multiple projects, while a project is part of the solution.

- **Location:** The location on your local computer where the files created to work with the project will be stored.

- **Solution name**: The name of the solution. In big projects, it is common to have multiple projects within a solution.

- **Place solution and project in the same directory**: Specifies whether the solution file (.sln) and project file (.proj) will be in the same directory or whether a solution file will be created as the root and a folder for the project:

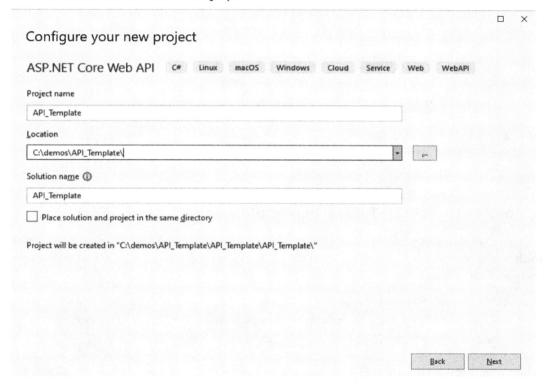

Figure 2.5 – Project configuration

Now that we know how to use the template selection window, let's look at the most important templates for .NET development.

Templates for web development

The first group we will analyze is the templates related to web development. To install these templates, with the VS Installer, you must select the **ASP.NET and web development** workload, as shown in *Figure 2.6*:

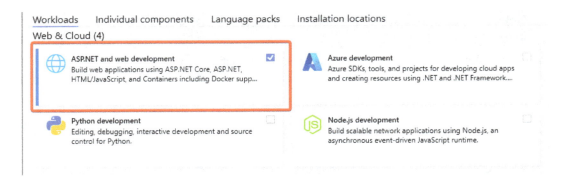

Figure 2.6 – Selecting the "ASP.NET and web development" workload

The description mentions that the templates of this workload allow you to create projects using ASP. NET Core, ASP.NET, HTML/JavaScript projects, and Containers.

Let's look at the main categories of templates that can be found with this workload next.

Templates for ASP.NET Core web applications

ASP.NET Core is a multiplatform web technology for creating modern applications using .NET, with which is possible to create standard web applications. The main templates that we can find to develop this type of application are as follows:

- **ASP.NET Core Web App (Razor Pages)**: This is the template for working with ASP.NET Core projects, using Razor Pages. It contains an example with ASP.NET Core Razor Page content.

- **ASP.NET Core Empty Project**: This is a blank template to place your own Razor pages.

To better understand how ASP.NET Core projects are built, let's do a test by selecting the **ASP.NET Core Web App (Razor Pages)** template. After creating a project with the selected template, you must fill out the information required for the new project, as shown in *Figure 2.7*:

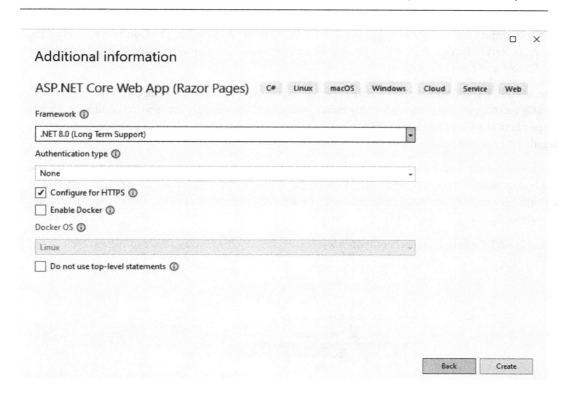

Figure 2.7 – Creating an ASP.NET Core Empty project

Here, you can configure the following options:

- **Framework**: This dropdown will show you the set of frameworks available to work with the selected technology. Although it is possible to select older frameworks, it is always advisable to create new projects with the latest version of the framework available.

- **Authentication type**: The authentication method to be used in the application.

- **Configure for HTTPS**: This checkbox allows you to configure the project to use a self-signed SSL certificate. If you are working on such a project for the first time, you will be asked to trust the certificate when you run the application so that everything works correctly. Although it is possible to work with the HTTP protocol, it is always recommended to use HTTPS in real life because it adds an extra layer of security by implementing the encryption between the user's browser and the web server, protecting sensitive data from possible external attacks.

- **Enable Docker**: This option allows you to enable Docker support in your project. This means that a Dockerfile will be generated, which you can then publish – for example, to Microsoft Azure.

- **Do not use top-level statements**: If this option is selected, a `Program` class and a `Main` method are created; otherwise, top-level statements are used.

In our example (*Figure 2.7*), we will select **.NET 8** as the target framework. The **Configure for HTTPS** option is marked by default, and it is optional for this demo. Finally, you can click on **Create** to complete the wizard and create the project.

The **ASP.NET Core Web App** template is perfect if you want to create a web application using .NET and C# running on the server. Also, this framework uses Razor pages (syntaxes that combine C# with **HyperText Markup Language** (**HTML**) in cshtml extension files) to build a web application into small and reusable pieces.

Modern applications normally run in the browser because they run faster and have better **search engine optimization** (**SEO**), which improves traffic to your website. However, server-side applications are still very useful for dashboards, internal projects, administration panels, and many other types of web applications.

An ASP.NET Core web project contains a wwwroot folder, which you can see in *Figure 2.8*:

Figure 2.8 – The structure of an ASP.NET web project

The wwwroot folder is associated with static files, such as CSS, images, and JavaScript files. There is also a folder called Pages that contains all the UI pieces, such as the home page and the error page.

Index.cshtml is an example where you can see C# code mixed with HTML code. The @ character allows you to use the C# code in the file. @model, for example, sets the model type to map values on the page.

To run or execute the application, we need to use the dark green arrow in the toolbar (known as the *play button*), which will start the project in the debugging mode, or the button with the light green arrow in the standard toolbar at the right (known as the *start without debugging button*), which will start the project without debugging:

Figure 2.9 – The play button to start a project in VS

In our example, I have selected the option with debugging enabled, but any of the options are fine. We just want to see how the project looks.

> **Important note**
>
> There are some useful shortcuts to start a project. You can use *F5* on your keyboard to run the project in the debugging mode, and *Ctrl + F5* to start the project without debugging.

You can see the result of the execution in the following figure:

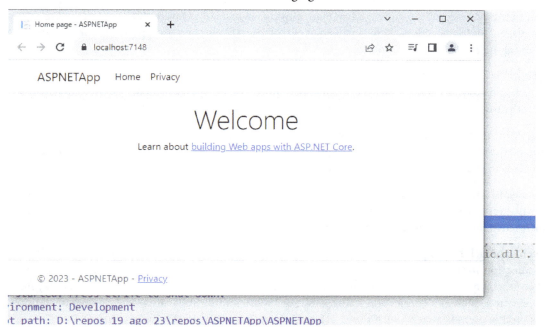

Figure 2.10 – Running a project with the ASP.NET Web Core template

> **Note**
>
> The topic of running and debugging applications is covered in detail in *Chapter 3, Debugging and Profiling Your Apps.*

Now, let's look at the templates for developing **single-page applications** (**SPAs**).

Templates for SPAs

An SPA is a type of application that consists of a single HTML page that loads in a user's browser. Interaction with users is based on JavaScript, which gives the feeling of being a more fluid and responsive application than a traditional web application.

SPAs are becoming increasingly popular as they provide a better performance, a smoother experience, and greater interactivity.

It is for this reason that VS includes a series of templates oriented to the creation of SPAs, which we will see next.

Blazor templates

Blazor is a framework released by Microsoft in 2019 that allows you to create native applications that run in the browser with the great advantage of using C# instead of JavaScript as the programming language. This framework has three main types of templates: **Blazor Server App**, **Blazor WebAssembly Standalone App**, and **Blazor Web App**.

Blazor WebAssembly Standalone App compiles C# code into code that runs in the browser locally. This means that the assemblies and resources required by the application are downloaded to the client's browser at startup so that once the resources are downloaded, the application runs smoothly.

On the other hand, **Blazor Server App** establishes a connection with a server that hosts the Blazor application, which prevents the entire application from being downloaded to the client, since the server is in charge of sending GUI updates to the browser.

Finally, **Blazor Web App**, Blazor's newest template, allows for the creation of static applications, or for the application to contain components that have a server-side interactive mode, a client-side interactive mode, or an **Auto** interactive mode.

Each approach has its pros and cons; for example, if an application created with Blazor Server loses connection to the server, it will not be able to continue running, while an application based on WebAssembly will be able to continue running. On the other hand, if you use sensitive data in your application, such as a connection string, a Blazor WebAssembly application will download all this data to the user's browser, which poses a risk, while in Blazor Server, this will not happen.

The templates for exclusive development with Blazor are as follows:

- **Blazor Server App:** This is the Blazor Server sample template, with pre-created components, such as the navigation menu, simple styles, added fonts, and other features.

- **Blazor WebAssembly StandAlone App:** This is the Blazor WebAssembly sample template, with pre-created components such as the navigation menu, simple styles, added fonts, and other features.

- **Blazor Server App Empty:** This template will create a project with the minimum files needed to start building the Blazor Server application from scratch.

- **Blazor WebAssembly App Empty:** This template will create a project with the minimum files needed to start building the Blazor WebAssembly application from scratch.

- **Blazor Web App:** In this template, it is possible to specify if the application will behave as a static application or if it will have components using the Blazor Server or Blazor WebAssembly model. Likewise, it is possible to specify whether the interaction mode will be defined globally or per component.

For this demonstration, I have selected the **Blazor WebAssembly StandAlone App** template, which will generate a project like the one shown in *Figure 2.11*:

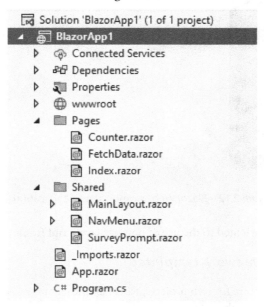

Figure 2.11 – The structure of a Blazor WebAssembly project

In this example, the Pages folder contains Blazor components intended to be pages in the application, while the Shared folder contains components that can be reused in various parts of the application.

Once we run the application, you will be presented with a browser window with a side menu, each option being one of the pages in the Pages folder, as shown in *Figure 2.12*:

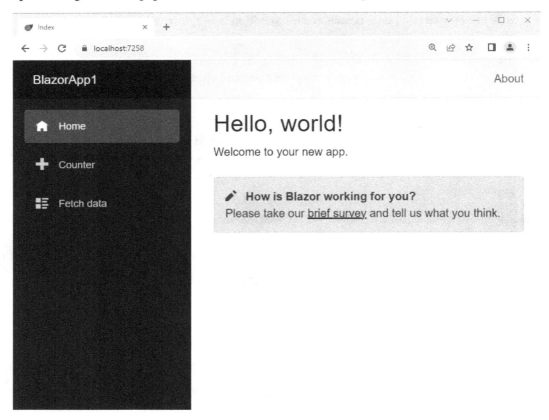

Figure 2.12 – Blazor WebAssembly demo application

Now, let's look at templates oriented to the use of popular JavaScript frameworks to create SPAs.

Popular JavaScript framework templates

SPAs are an amazing architecture for web projects, where all the elements are rendered using a single HTML file. There are a good number of libraries and frameworks that use this concept – for example, Angular, React.js, and Blazor WebAssembly.

Any project created with JavaScript framework templates will contain all the required components to create a monolithic application, using ASP.NET on the backend and an SPA library or framework on the frontend side.

Now, let's create our first SPA project with VS. You can search by entering and `ASP.NET Core` to find templates for SPAs using JavaScript:

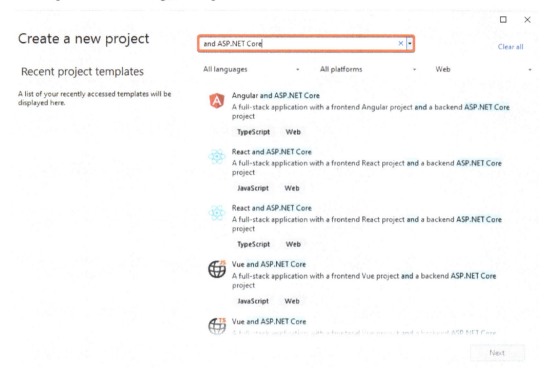

Figure 2.13 – Searching by entering "and ASP.NET Core"

In *Figure 2.13*, you can see the result of searching by entering and `ASP.NET Core`. There are three main templates that we can select to work with SPAs:

- **Angular and ASP.NET Core**
- **React and ASP.NET Core** (JavaScript and TypeScript version)
- **Vue and ASP.NET Core** (JavaScript and TypeScript version)

Let's do a test by selecting the **React and ASP.NET Core** template.

Once you have indicated the name – in my case, `SPAProject` – the path where the project will be saved, and the target framework, you can see the project created with the SPA template and analyze the architecture:

Figure 2.14 – The ASP.NET Core project with the React project created in VS

There are three important folders to highlight in the **React and ASP.NET Core** template:

- `ClientApp`: Contains the client application – in this case, a React.js app
- `Controllers`: Contains all the controllers related to the business logic on a server
- `Pages`: Contains Razor pages, which means UI components rendered on a server

The template has a demo with the `WeatherForecastController.cs` file. This is a simple demo that returns some random data.

Let's run the project by clicking the play button, as shown in *Figure 2.15*, to see how the application looks:

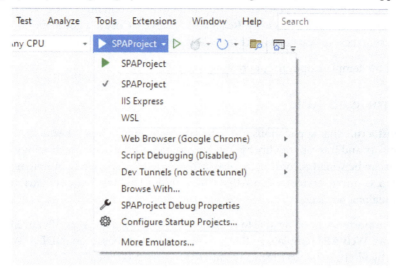

Figure 2.15 – The option to run the project in VS (the play button)

When the project is running, you can see a simple HTML page explaining how the template works, which includes two demos, **Counter** and **Fetch data**:

Figure 2.16 – ASP.NET Core with the React.js project running

The template is simple but includes everything that we need for creating a new web application using a monolithic architecture and best practices on the backend and frontend sides. If you require a web application with high-performance security, **React and ASP.NET Core** is a good option. React.js uses JavaScript in the syntaxes, so you need more knowledge of this language and C# for the backend.

Now, let's look at the template options for creating web APIs.

Templates for web APIs

Today, it is almost a rule that applications use API-based endpoints. This is because APIs provide a high level of security and interoperability by not depending on a particular technology or operating system, and they can be scaled according to the existing demand. Also, we can implement advanced architecture, such as microservices using APIs, where all our business logic is distributed in small isolated and standalone services.

Due to the great importance of being able to develop solutions based on APIs, VS 2022 incorporates the **ASP.NET Core Web API** template so that we can create APIs based on .NET 8. We can search for this template by filtering the `api` term, as shown in *Figure 2.17*:

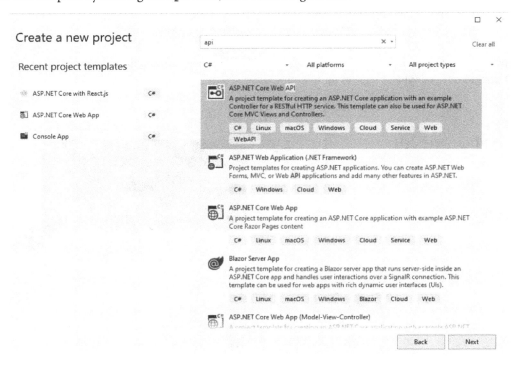

Figure 2.17 – Filtering by api and selecting the ASP.NET Core Web API

Once we have selected the template, a new window will appear, in which we will be asked to select the version of the framework and whether we require some type of authentication, among other data that we have already seen in the *Templates for ASP.NET Core web applications* subsection. However, we have a couple of additional options that we do not see in a normal ASP.NET Core project, which are as follows:

- **User controllers**: If this option is selected, the use of controller files will be enabled from the configuration. If it is deselected, a feature called minimal APIs will be used that will create the minimum code necessary to have a functional API.

- **Enable OpenAPI support**: Swagger is a set of open source tools based on the **OpenAPI Specification** (**OAS**) that will allow us to describe the APIs we create in a simple, easy, and well-structured way, providing API users with good documentation.

In my case, I selected **.NET 8.0 (Long Term Support)** as the framework, an **Authentication type** value of **None**, and checked the **Configure for HTTPS** option in the **Additional information** section, as shown in *Figure 2.18*:

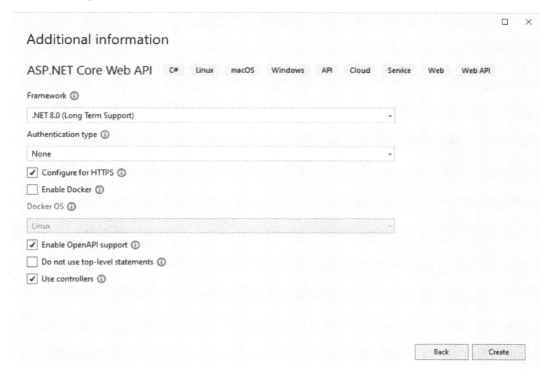

Figure 2.18 – Selecting options in "Additional information" for the ASP.NET Core Web API project

Once the project has been created, we can see that a folder called `Controllers` has been created as part of the project:

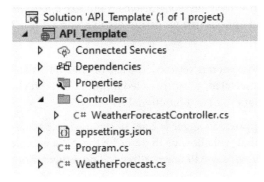

Figure 2.19 – The structure of an ASP.NET Core API project

In the `Controllers` folder, we will find the controllers that will be created as part of the API. Remember that we can see the controllers in this folder because of the **Use controllers** option we selected in *Figure 2.18*.

If we proceed to run the application, it will immediately take us to a window with a URL ending with `/swagger/index.html`:

Figure 2.20 – The Swagger page for the ASP.NET Web API project

Figure 2.20 shows the Swagger UI tool, which allows us to test the endpoints of our project interactively and quickly. Here, we will be able to find all the functionalities available in our API documented through Swagger. Swagger is a set of open source tools (including Swagger UI) that uses the OAS to document the endpoints of the controllers found in a project, including the parameters required and the types of values returned.

From the same page, we will be able to test each one of the endpoints, for the purpose of validating them, and carry out necessary debugging when some endpoint does not work as expected. As you can appreciate, Swagger is of great help both for development purposes and for providing users with documentation of the created API.

ASP.NET Core Web API gives us everything we need to create our own modern APIs from scratch.

Let us now look at the available options that will allow us to create multiplatform applications.

Templates for multiplatform development

.NET MAUI is Microsoft's framework that allows the development of multiplatform applications – that is, applications for Android, iOS, and Windows mainly, although it is possible to deploy applications on other platforms such as Tizen and macOS.

In order to be able to use these templates, in the Visual Studio Installer, we must select the **.NET MAUI Multi-platform App UI development** workload, as shown in *Figure 2.21*:

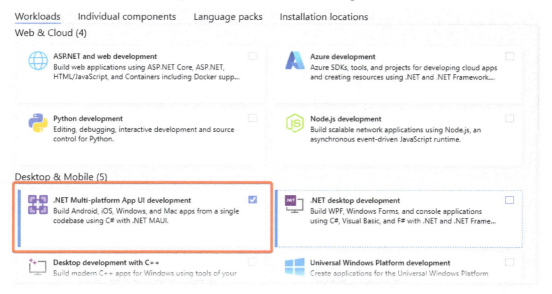

Figure 2.21 – The ".NET MAUI Multi-platform App UI development" workload

This type of project uses markup files with a `.xaml` extension for the definition of the graphical interface while C# code is used to specify the business logic.

For example, to display text in a `.xaml` file, a syntax like the following is used:

```
<Label
    FontAttributes="Bold"
    FontSize="30"
    Text="Packt"
    TextColor="White" />
```

This allows us to define the design of the pages of an application for every platform in `.xaml` files. With this, we will avoid creating GUI files for each of the platforms, saving time and allowing better organization.

In VS, we can select one of the two available templates to create applications with .NET MAUI:

.NET MAUI App: This is the ideal template if you want to create a .NET MAUI application from scratch. The project has a single XAML page called `MainPage.xaml`, and all the files and folders necessary for the development of your application.

.NET MAUI Blazor App: This template allows you to create an application that can access native features of the platforms, with the advantage of being able to use the same Blazor components that you could have used for an SPA or if you are a specialist in the use of HTML and CSS. Even with the new version of .NET 8, it is possible that these components are JavaScript components using React, for example.

Once you select the default .NET MAUI template, you will see a structure like the one in *Figure 2.22*:

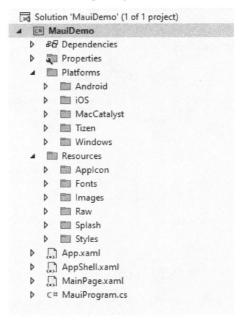

Figure 2.22 – Structure of a .NET MAUI project

The most important thing to know about these templates is that they contain a folder called Platforms, which contains subfolders with the name of each platform on which we can deploy the application (such as Android, iOS, and Windows) that contain platform-specific files. It is possible to modify these files or add additional files to affect the behavior of the application on a specific platform.

Another welcome feature is a folder called Resources, also located in the root directory, which contains subfolders with common resources that will be used throughout the application, such as Images, Fonts, Styles, and so on. This makes it much easier to manage resources since they are centralized in one place. In addition, .NET MAUI takes care of complex tasks, such as .svg image resizing, splash screen image handling, and so on.

To deploy a .NET MAUI project, being a multiplatform project, emulators must be created in the case of Android, a simulator must be used for iOS, and the developer mode must be enabled in the case of Windows. In all these cases, it is also possible to test directly on physical devices. Unless a special code has been written for any of the platforms, the result of the GUI should be similar for all of them, as shown in *Figure 2.23* and *Figure 2.24*.

In *Figure 2.23*, we can see the implementation of the demo application on an Android emulator:

Figure 2.23 – The demo application running on an Android emulator

In *Figure 2.24*, we see the same demo application but now deployed as a Windows application. You can see that the design is the same, although the platform is different:

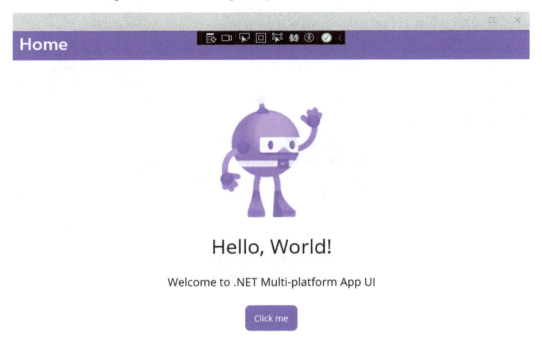

Figure 2.24 – The demo application running on a Windows machine

Now that you know about the templates for .NET MAUI development, let's look at the templates for desktop development.

Templates for desktop development

The last category of templates we will look at are desktop development templates. To install these templates on your computer, you must select the workload called **.NET desktop development**, as shown in *Figure 2.25*:

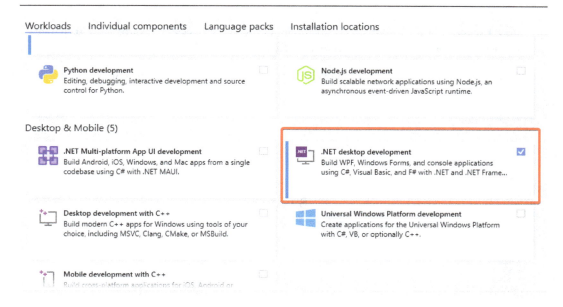

Figure 2.25 – The ".NET desktop development" workload

This workload includes templates for two technologies with many years of maturity: **Windows Forms** and **Windows Presentation Foundation (WPF)**.

Windows Forms is an old but still widely used technology among .NET developers, since it allows us to create desktop stacks quickly thanks to its drag-and-drop solution. The main templates for Windows Forms development include the following:

- **Window Forms App**: This is the base template for creating a new Windows Forms application. It includes a file so that you can begin to design your first window quickly.

- **Windows Forms Control Library**: A project that allows us to manage customized controls for our applications, which can be referenced by Windows Forms applications.

On the other hand, *WPF* is a technology that uses an alternate version of the XAML language used in the .NET MAUI platform for the creation of graphical interfaces, only oriented exclusively to desktop applications. Although there are similarities between WPF and .NET MAUI, not all XAML code can be reused between both platforms, due to the difference in properties and controls.

WPF, as with Windows Forms, allows the design of graphical interfaces with drag-and-drop windows, besides having more customization options for each control, making it the preferred option for Windows desktop development. Among the templates that we can select to work with WPF are the following:

- **WPF Application**: This template is the one you should select to create your WPF applications. It creates a blank window so that you can start immediately with your development.

- **WPF Custom Class Library**: This template allows you to create custom controls and keep them as a separate library so that you can reuse them in different WPF projects.

For a quick demonstration, I will select the **WPF Application** template to create a new project:

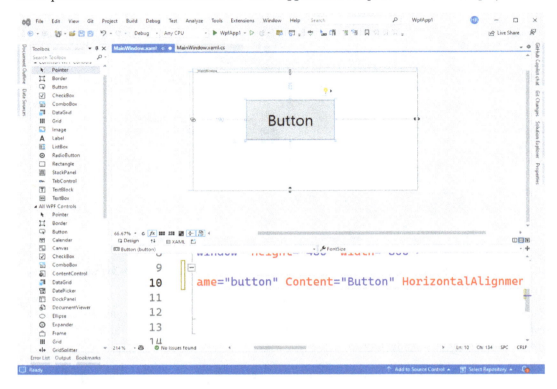

Figure 2.26 – WPF designer

As shown in *Figure 2.26*, VS splits the screen to be able to work either with the window designer that looks like a blank canvas or with the XAML code at the bottom, which is a great advantage while creating applications. Practically, what you see in the designer will be the result of the running application.

In *Figure 2.27*, you can see the comparison between a WPF application on the left and a Windows Forms application on the right with standard controls, in which I have changed some simple properties:

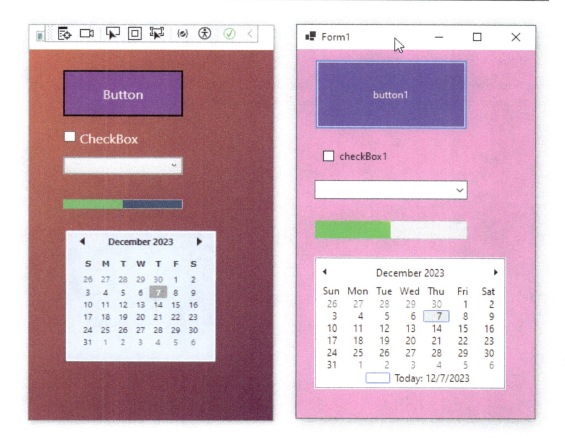

Figure 2.27 – WPF application on the left and Windows Forms application on the right

As you can see, the WPF application looks aesthetically better since it is possible to style the controls with greater flexibility thanks to the use of the XAML language.

Summary

In this chapter, we have seen the main workloads and the most common templates available in VS for developing different types of projects.

We have seen specific templates for web development, for creating ASP.NET Core applications, SPAs, and web APIs. Undoubtedly, ASP.NET and web development is one of the workloads that has more templates available, because VS is not only designed for .NET developers but also for web developers in general.

We also analyzed templates that allow cross-platform development with .NET MAUI, allowing the deployment of applications for Android, iOS, and Windows mainly. Finally, we saw which templates are available for desktop development.

In the next chapter, you will learn about tools available in VS that will allow you to debug your applications to detect errors, as well as tools to analyze the performance of your applications.

3

Debugging and Profiling Your Apps

As software developers, one skill that should be learned as early as possible is program debugging. This applies to .NET projects as well as to any other software development technology.

If you want to get the most out of VS 2022, you must be familiar with its different windows, which can help you observe information so that you can fix bugs and know how to use as many of the debugging tools it offers, including **breakpoints**. A breakpoint offers the functionality to stop the execution of an application, allowing you to see the state of each of the objects and corroborate its behavior.

Another set of tools that will help you on your way as a professional developer is profiling tools, which help with finding bottlenecks that can cause poor application performance.

That is why, in this chapter, we will talk about the different tools and options for debugging and profiling applications in VS 2022.

The topics we will discuss in this chapter are as follows:

- Understanding compilation in VS 2022
- Building a project in VS 2022
- Debugging projects in VS
- Exploring breakpoints in VS
- Inspection tools for debugging
- Measuring app performance with profiling tools

Let's learn about debugging and profiling in VS 2022 so that we can detect possible errors and underperformance in our programs.

Technical requirements

To follow the examples presented in this chapter, VS 2022 must be installed with any workload, as described in *Chapter 1, Getting Started with Visual Studio 2022*.

You can find the practice project for this chapter at `https://github.com/PacktPublishing/ Hands-On-Visual-Studio-2022-Second-Edition/tree/main/Chapter%203`.

> **Important note**
> VS has keyboard shortcuts for almost all the operations we are going to be performing. So, I will mention them for you as *Shortcut* notes as we continue with this chapter.

Understanding compilation in VS 2022

VS is an excellent IDE that makes the process of compiling applications transparent to developers. For example, after creating a project, it is enough to press a button to run any application. However, understanding some details of the VS compilation process can help you optimize code, diagnose errors, and take full advantage of the IDE.

In this section, I am not going to go into the details of concepts such as what a compiler is or the process of converting high-level code into machine language; instead, I am going to give you some tips and advice that you can take advantage of.

Let's look at how we can compile applications in VS.

How to build a project in VS 2022

Once you create a project in VS using any of the templates we discussed in *Chapter 2, Creating Projects and Templates*, one of the first things you should do is compile the project. This is because, during the steps of compilation, VS converts the high-level code into a machine language, while also taking care of managing the project dependencies, such as the associated NuGet packages, to make all the necessary files and libraries available for correct compilation.

There are several ways to carry out compilation in VS: one of them is to go to the **Solution Explorer**, then position ourselves on the name of the project we want to compile and right-click. This will show us three options – **Build**, **Rebuild**, and **Clean** – that will be very useful during our lives as .NET developers as they will allow us to start the process of compiling or cleaning projects and thus be able to execute them:

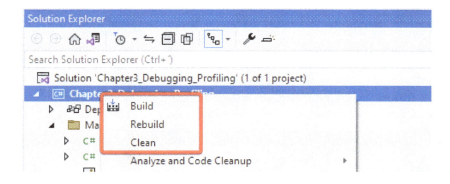

Figure 3.1 – The Build, Rebuild, and Clean options in VS

It is also possible to use keyboard shortcuts to carry out **Build**, **Rebuild**, and **Clean** tasks. This will be detailed in *Chapter 13*, *Learning Keyboard Shortcuts*.

Understanding the difference between these options is essential to determine when to use each of them, so we will discuss them next.

Build versus Rebuild versus Clean

In this section, we'll learn about what the three options that are available during compilation (namely, **Build**, **Clean**, and **Rebuild**) do:

- **Build**: First, the **Build** operation will perform a build on files that have been modified since the last build was performed. That is why, after performing the first build and run, you will notice a considerable speed increase when you test an application again. **Build** is the most used option since it allows faster development by enabling fast compilations.

- **Clean**: There are times when you will want to remove all binary and temporary files generated during a build, such as when you want to share or archive source code without executables or files that turn your folders into space-consuming monsters. This is when the **Clean** option is the best choice. This option only cleans, so it does not compile any files.

- **Rebuild**: Finally, the **Rebuild** option is a combination of the **Clean** and **Build** operations. First, it removes all the binary files generated in the last compilation and then compiles the project from scratch. This option is useful when there may be residual files that are causing problems in the compilation, or when you are facing unexplained errors that cannot be solved with normal compilation since the compilation is started from a clean base.

Now that you know the compilation operations that are available in VS, let's see what the result of the compilation is.

What happens after compilation?

After starting the compilation process, at the bottom of VS, you will see a window called **Output**. If you want to see information about the compilation process, you can select the **Build** option, as seen in *Figure 3.2*, which will display information about the compilation process that was performed. The number of lines that are displayed will depend on the complexity of the compilation; for example, the more files and components a project has, the more complex the process will be:

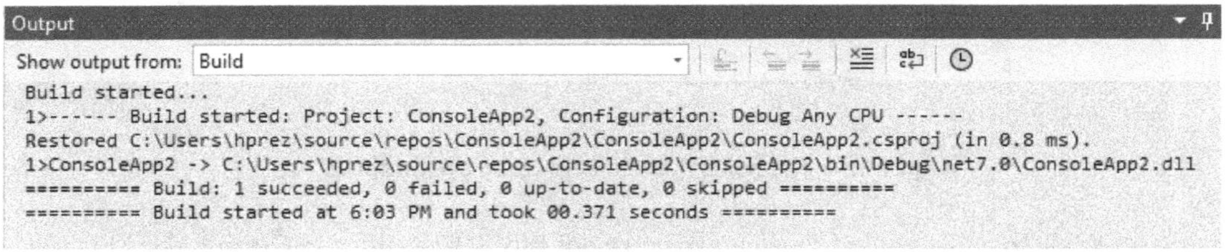

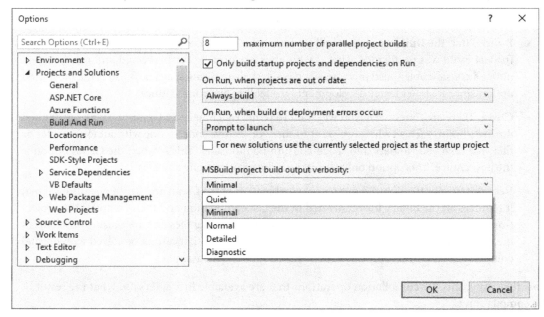

Figure 3.2 – The Output window

There are other cases, such as a newly created and compiled console project, that will show you just a couple of lines, as shown in *Figure 3.2*, but if you want to know what has happened in detail, you can view the step-by-step process by navigating to **Tools | Options | Projects and Solutions | Build And Run**. Here, you must change the value of **MSBuild project build output verbosity**, which has a value of **Minimal** by default, as shown in *Figure 3.3*:

Figure 3.3 – Options for displaying compilation details in the Output window

By changing the value of **MSBuild project build output verbosity** to **Detailed**, you will see the difference, as shown in *Figure 3.4*:

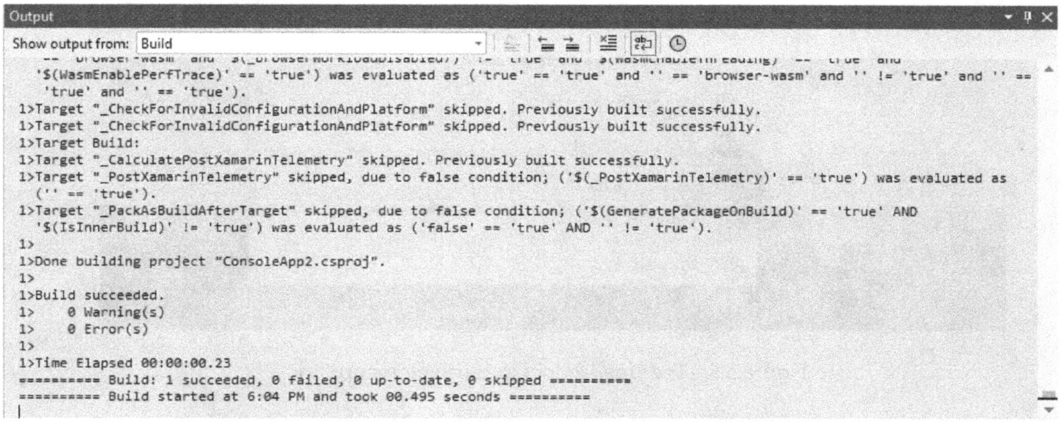

Figure 3.4 – The Output window with a detailed compilation report

Let's take a brief look at the types of messages that will be displayed for each type of output verbosity:

- **Quiet**: Errors and warnings

- **Minimal**: Errors, warnings, and high-importance messages

- **Normal**: Errors, warnings, high-importance messages, and normal-importance messages

- **Detailed**: Errors, warnings, high-importance messages, normal-importance messages, and low-importance messages

- **Diagnostic**: Errors, warnings, high-importance messages, normal-importance messages, low-importance messages, and additional MSBuild engine information

Another thing you should know is that once the compilation is finished, several folders and files are created as a result. If you want to see what these are, you can position your mouse over the project, right-click, and select the **Open Folder in File Explorer** option:

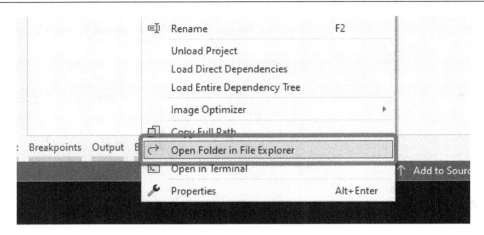

Figure 3.5 – The Open Folder in File Explorer option

This option will show you the different folders and files generated by the compilation with a hierarchy. The following one has resulted from my compilation:

```
...\source\Chapter3_Debugging_Profiling\Chapter3_Debugging_Profiling\
bin\Debug\net8.0
```

In this hierarchy, inside the `bin` folder, you will find the execution configuration that was used, as we will discuss in the *Debugging projects in VS* section; this is followed by a folder that specifies the version of the framework used, inside which we'll find the executable files that were used to deploy the application – in our example, a console application labeled in the same way as how we named the project (in my case, `Chapter3_Debugging_Profiling.exe`), as shown in *Figure 3.6*:

Name	Date modified	Type	Size
Chapter3_Debugging_Profiling.deps.json	8/18/2023 5:08 PM	JSON File	1 KB
Chapter3_Debugging_Profiling.dll	8/18/2023 5:08 PM	Application exten...	10 KB
Chapter3_Debugging_Profiling.exe	8/18/2023 5:08 PM	Application	152 KB
Chapter3_Debugging_Profiling.pdb	8/18/2023 5:08 PM	Program Debug D...	13 KB
Chapter3_Debugging_Profiling.runtimec...	8/18/2023 5:08 PM	JSON File	1 KB

Figure 3.6 – Files generated by a compilation

Now that you have learned about the compilation process, it is time for you to learn how to debug applications in VS 2022.

Debugging projects in VS

Before discussing breakpoints in depth in the *Exploring breakpoints in VS 2022* section, you must know about some technical aspects that are used in the debugging world, as well as in VS.

We'll start with what *debugging* and *debugger* are.

Understanding the technical aspects of debugging

You must know the difference between the terms **debugger** and **debugging** so that you know what I mean when I mention any of these terms throughout this book.

First, the term *debugging* refers to the action of looking for errors in the code. This does not necessarily include the use of a tool such as an IDE. You could, for example, search for errors in code written on a piece of paper, and you would still be debugging.

This is usually not feasible, and a tool called a *debugger* is often used. This tool is attached to the application process you are going to run, allowing you to analyze your code while the application is running.

In VS, we can start an application with the debugger or without it. In the next section, we will learn how to configure a project in both modes.

Differentiating between debug mode and release mode

It is essential to differentiate between debug mode and release mode in VS as they can be confusing to those who touch the IDE for the first time.

First, let's analyze debug mode. This option can be activated by selecting the **Debug** configuration (which is pre-selected by default) and clicking on the green button located in the same space as the project's name, as shown in *Figure 3.7*:

Figure 3.7 – VS's Debug mode option

When this option is selected, the debugger will be attached to the execution of the application, which will allow us to use functions, such as stopping at a certain breakpoint in our application.

On the other hand, we can also choose a second configuration from the drop-down list, as shown in *Figure 3.8*, called **Release**:

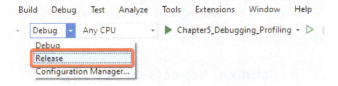

Figure 3.8 – VS's Release mode option

When this option is selected and you start the execution of the application, the debugger will not be attached, which will give you a better idea of how your application will behave toward the end user. This implies that you will not be able to perform code debugging or see where exceptions have occurred, but you will achieve performance improvements and the generated binaries will be smaller in size.

Project debugging initialization options

As part of the debugging and execution options of an application, we must know that we have a set of options available to perform our tests.

If you drop down the options next to the green button specifying the name of your project, as shown in *Figure 3.9*, you will be able to see a set of configurations for the deployment of your application:

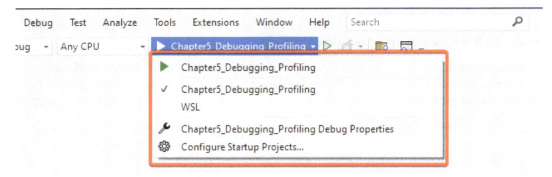

Figure 3.9 – The application configuration options for debugging

For example, in the console application, it is possible to test an application on Windows with the default option selected, or even **WSL**, if your application is more oriented to Linux-based environments. In the case of other types of projects, such as web or mobile, you can choose which browser you wish to use or the emulator where the application will be executed, respectively.

If the application has been configured in **Debug** mode, when launched, you will see the buttons that allow you to control the execution of the application in the upper part of the IDE, as shown in *Figure 3.10*. These buttons, from left to right, are used to do the following:

- Pause the application

- Stop the application

- Restart the application:

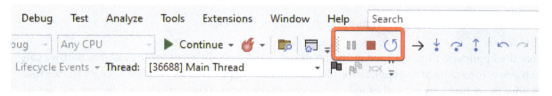

Figure 3.10 – Buttons to control the flow of the application

Now that we know about the existing debugging concepts in VS, let's analyze breakpoints.

Exploring breakpoints in VS

Breakpoints are a fundamental part of software development. They allow you to stop the flow of your application at any point where you want to inspect the state of your objects.

To place a breakpoint in VS, it is enough to position ourselves right next to the numbering of the lines. A gray circle will appear and disappear as we move our cursor over the line numbers.

Once we find the line we want to debug, we just need to left-click once, which will cause the circle to turn red, as shown in *Figure 3.11*. Once you have done this, if you move the cursor away from the circle, you will be able to see that it remains as-is, as shown in the following screenshot:

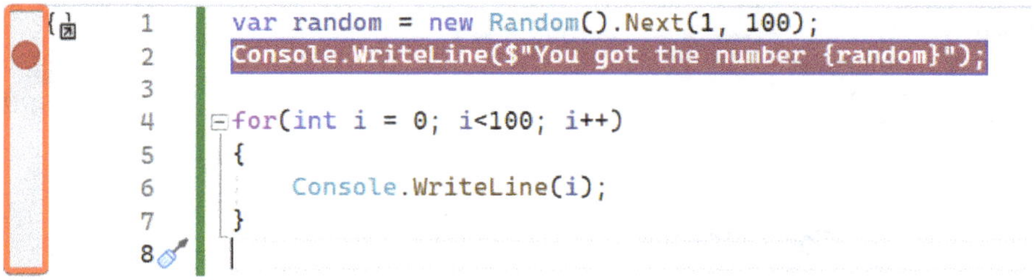

```
1    var random = new Random().Next(1, 100);
2    Console.WriteLine($"You got the number {random}");
3
4    for(int i = 0; i<100; i++)
5    {
6        Console.WriteLine(i);
7    }
8
```

Figure 3.11 – Placing a breakpoint

If we proceed to execute the application with the breakpoint set, we will see how the application flow stops immediately after starting the application, as shown in *Figure 3.12*:

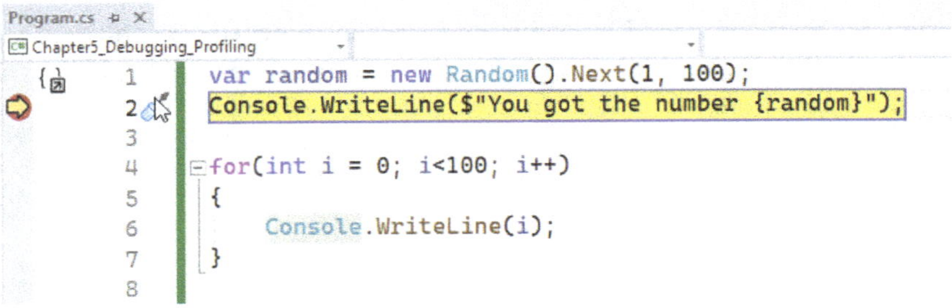

Figure 3.12 – Debugging a breakpoint

Once the application has stopped at the breakpoint, we have different tools available that we can use to view the status of the application; for example, if we position our cursor over a variable that is before the debug line, we will be able to see its current status. If it is primitive data, you will see its value immediately, while if it is an object, you will be able to go inside its properties to examine each one of them, as shown in *Figure 3.13*:

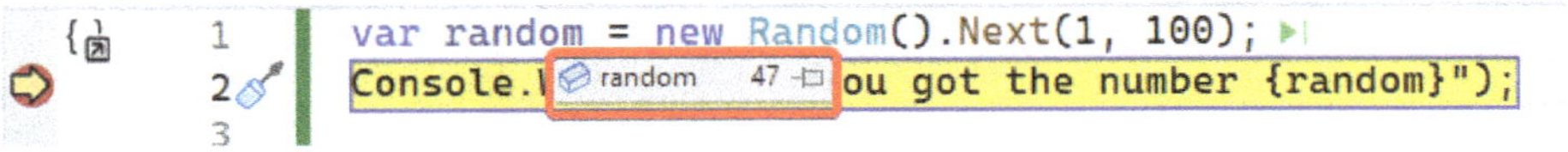

Figure 3.13 – Examining the properties of an object

There are more options for placing breakpoints; for example, if you place your cursor over one of the gray circles and right-click on it, a series of breakpoints will appear, as shown in *Figure 3.14* (we will analyze these in the *Conditional breakpoints*, *Temporary breakpoints*, and *Dependent breakpoints* sections), so that you can insert them:

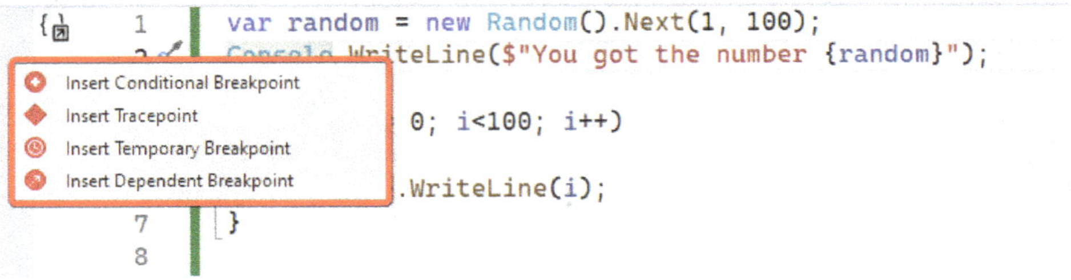

Figure 3.14 – The options for inserting a breakpoint

Likewise, if you right-click on any breakpoint that's already been placed, you will see options to add functionalities to it, as shown in *Figure 3.15*:

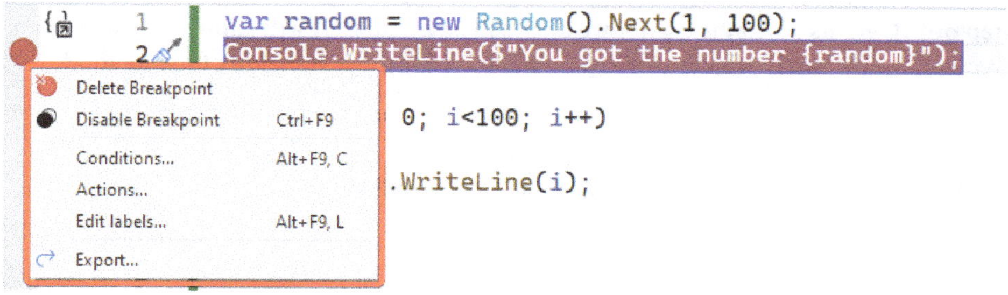

```
{ᵇ      1      var random = new Random().Next(1, 100);
        2      Console.WriteLine($"You got the number {random}");
```

Delete Breakpoint
Disable Breakpoint   Ctrl+F9   0; i<100; i++)
Conditions...   Alt+F9, C
Actions...   .WriteLine(i);
Edit labels...   Alt+F9, L
Export...

Figure 3.15 – The menu for modifying breakpoints

This is the easiest way to add breakpoints to your project. However, you will often need special breakpoints that are activated under certain circumstances. We will review these in the *Different types of breakpoints* subsection.

> **Shortcut**
>
> It is possible to enter a breakpoint quickly by placing your cursor on the line you want to debug and pressing the *F9* key.

Navigating between breakpoints

Once we know how to place breakpoints in the source code, we can continue executing the application in different ways through the buttons located in the upper part of the menu:

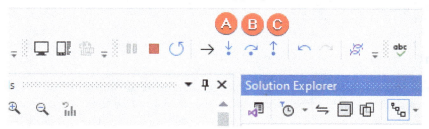

Figure 3.16 – The debugging options for application execution

Each of these buttons (from left to right in *Figure 3.16*) will execute the code as follows:

- **Step into** (*A*): This allows us to execute all the source code line by line. If, as part of the flow, we invoke methods to which we have access through the source code, we will navigate to it automatically while following the line-by-line debugging process.

- **Step over** (*B*): This allows us to only walk through lines of code in the current method and not step into any methods that are invoked by the current method.

- **Step out** (*C*): This button is used if we are inside a method. It will allow us to step out of the execution of the method to return just to the line after the invocation of the method.

Shortcut

Each of the options in the preceding list has a shortcut assigned to it:

Step into: The *F11* key

Step over: The *F10* key

Step out: The *Shift* + *F11* keys

If we wish to examine all the breakpoints we have in our project, we can do so by activating the **Breakpoints** window. To activate this window, navigate to **Debug | Windows | Breakpoints**, as shown in *Figure 3.17*:

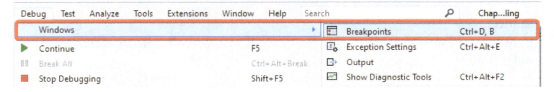

Figure 3.17 – The option to display the Breakpoints window

This will display a new window that shows a list of all the breakpoints that have been placed in our project.

Now that you know the general concepts of breakpoints, let's analyze the different types of breakpoints available in VS.

Different types of breakpoints

In VS, we have multiple types of breakpoints that can help us debug our code faster, for example, by defining which groups of breakpoints must be activated at a certain time. You must know about these breakpoints since they will be part of your toolbox when you're creating code. So, let's study the different types of breakpoints.

Conditional breakpoints

There may be times when you need your breakpoint to stop when certain conditions are met. In this case, using **conditional breakpoints** is the best option. To insert a conditional breakpoint, just right-click on the sidebar; you will be shown the different types of breakpoints that are available, as described in the *Navigating between breakpoints* section. Select the **Conditional Breakpoint** type to open a window that provides preselected options, as shown in *Figure 3.18*:

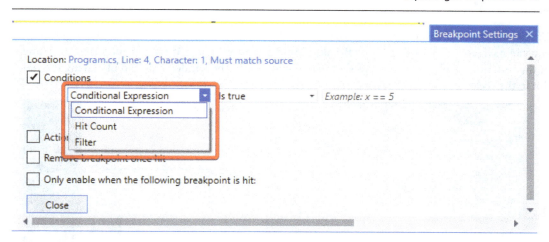

Figure 3.18 – The options for conditional breakpoints

Within a conditional breakpoint, we can configure either a **Conditional Expression** condition, a **Hit Count** condition, or a **Filter** condition. Let's see under which conditions should we use these options:

- **Conditional Expression**: This allows us to stop the application when a true condition that we have previously specified is fulfilled, or when the value of an object changes.

- **Hit Count**: This is ideal to be used in cycles. It allows us to specify a comparison that, when fulfilled, stops the application at the specified point. For example, if, in a cycle of 100 iterations, starting with index 0, the condition is specified to be equal to 50, the app will stop when the index is equal to 49.

- **Filter**: This will allow us to trigger a breakpoint according to a series of predefined expressions ranging from the machine name to processes, to thread properties.

> **Shortcut**
> To insert a conditional expression, you can use the *Alt + F9* keys, followed by the *C* key.

Now, let's take a look at function breakpoints.

Function breakpoints

The **Function Breakpoint** type, as its name implies, will allow us to debug a method when it is executed, even if we have not set a breakpoint, as we did in the *Conditional breakpoints* section. A **Function Breakpoint** type is very useful if you have hundreds of lines of code and know the name of the function that you want to debug.

Unlike conditional breakpoints, functional breakpoints are placed differently. First, as seen in *Figure 3.19*, you can navigate to the **Debug | New Breakpoint | Function Breakpoint** menu and insert the name of the function in which we want to set the breakpoint into the window, instead of placing the red dot in the code:

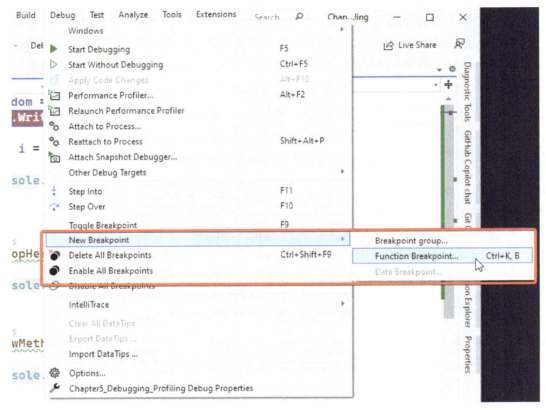

Figure 3.19 – Adding a function breakpoint from the Debug menu

Another way is to add a breakpoint function from the **Breakpoints** window, as shown in *Figure 3.20*:

Figure 3.20 – Adding a function breakpoint from the Breakpoints window

Once we click on the **Function Breakpoint** button, a new window will pop up, asking us for the name of the method we want to monitor. For this demonstration, we have introduced a very simple function named `StopHere` to the `Program.cs` file that's in the following format:

```
void StopHere()
{
    Console.WriteLine("Hi!");
}
```

Knowing the name of the function to be evaluated, we can enter it in the **Function Name** box. We can do this in several ways:

- Typing the name of the function

- Specifying the function name with a specific overload

- Specifying the `dll` name if we have the source code for it

In our example, we will only place the name of the `StopHere` function:

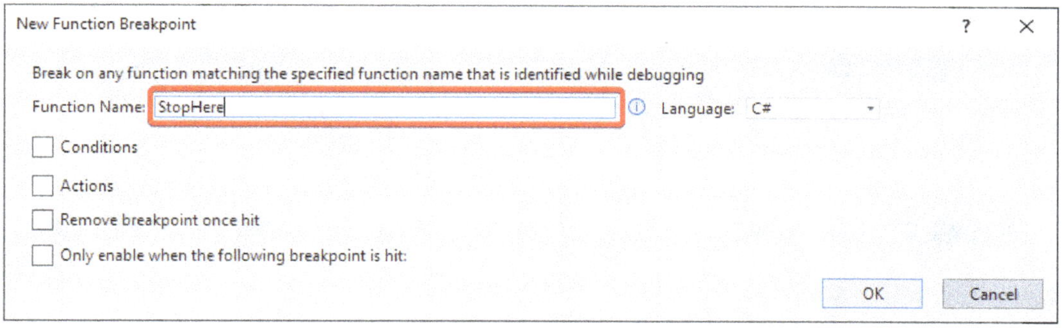

Figure 3.21 – Filling in the name of the function we are interested in debugging

With the data established, we will proceed to execute the application, having previously called this new method. This will cause the application to stop at the start of the method (which we specified in *Figure 3.21*), as seen in *Figure 3.22*:

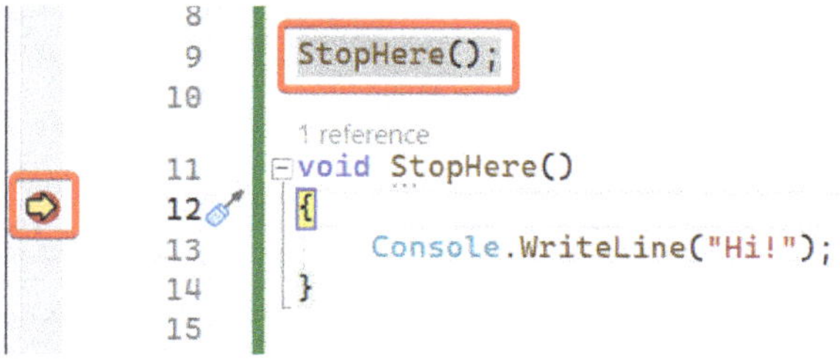

Figure 3.22 – A demonstration of the breakpoint being executed through a function breakpoint

> **Shortcut**
>
> It is possible to add a function breakpoint by pressing the *Ctrl + K* keys, followed by the *B* key.

Undoubtedly, a functional breakpoint will make our lives easier when we want to debug methods quickly. But in case you want to trigger breakpoints based on the data of an object, you can use the data breakpoints. We'll cover these next.

Data breakpoints

If you want to be able to place breakpoints when the properties of an object change, then **data breakpoints** are your best option. If you try to add one of these breakpoints from the **Breakpoints** window, you will see that the option is disabled.

This is because first, we need to place a breakpoint at a point before we want to start monitoring the property. Once we've done this, we can start the application until the breakpoint is activated and, in the **Locals** window, where the instance that interests us appears, right-click to see the **Break When Value Changes** option, as shown in *Figure 3.23*. Here, we want to monitor the Name property of the person object:

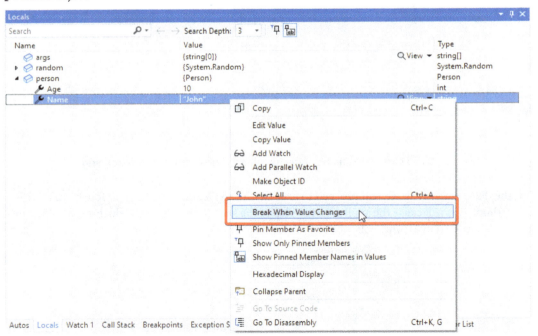

Figure 3.23 – The option to add a breakpoint when the Name property changes value

Once we select the option, we will see how a new breakpoint is created automatically. This will detect when the property we have specified changes, as shown in *Figure 3.24*:

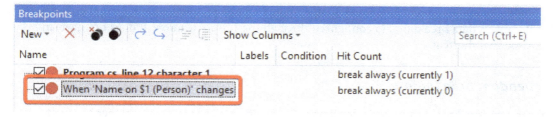

Figure 3.24 – The data breakpoint being created in the Breakpoints window

Let's say that we proceed to modify the property data – for example, through the following code:

```
Console.WriteLine(person.Name);
person.Name = "Peter";
```

Now, when we run the application, we will see how the breakpoint we placed previously is reached, as shown in the following screenshot:

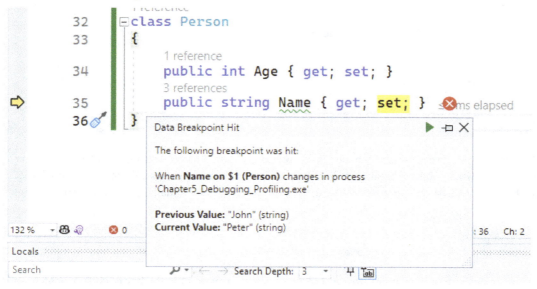

Figure 3.25 – A demonstration of a break when the property has been changed

Finally, you will see that when you restart the application or stop it, the breakpoint will disappear from the window. This is because the object reference has been lost.

Now, let's look at a new type of breakpoint that was introduced in VS 2022, called dependent breakpoints.

> **Note**
> Because a data breakpoint is configured through the **Locals** window, it does not have an associated shortcut.

Dependent breakpoints

Dependent breakpoints are special breakpoints that will only be executed if another breakpoint is reached first. Perhaps, in a simple scenario, they are not very useful, but in complex scenarios, where hundreds of functionalities come into play, they can be of great help.

For example, imagine that you have a method that is invoked in several places in your application, and you are testing a new functionality that invokes it. If you were to place a normal breakpoint on the method, it would stop every time it is invoked. With a dependent breakpoint, you can specify that you only want to stop execution if the breakpoint of your new functionality is reached.

To demonstrate this functionality, I have added a couple of methods to the Program.cs file, as follows:

```
void NewMethod()
{
    Console.WriteLine("New Method");
}

void CommonMethod(string message)
{
    Console.WriteLine(message);
}
```

The idea of the CommonMethod function is that we can see the content of a string passed as a parameter and know at what point the breakpoint has stopped. With this in mind, let's add some additional lines in which we will first call CommonMethod, then NewMethod, and, finally, CommonMethod again:

```
CommonMethod("Before invocation of NewMethod()");
NewMethod();
CommonMethod("After invocation of NewMethod()");
```

To place a dependent breakpoint, we must place a normal breakpoint in the line of code on which we want to depend – that is, the one that must be executed for the dependent breakpoint to be executed. In our example, we will place a breakpoint without any configuration in the NewMethod functionality since it is the method we want to test, as shown in *Figure 3.26*:

```
           1 reference
26    ⊟void NewMethod()
27     {
28         Console.WriteLine("New Method");
29     }
30
```

Figure 3.26 – The placement of the breakpoint on which a dependent breakpoint will depend

Then, you must right-click on the line where you want to create the dependent breakpoint, as shown in *Figure 3.27*:

```
           2 references
31    ⊟void CommonMethod(string message)
32     {
33⟋        Console.WriteLine(message);
```
   ⊕ Insert Conditional Breakpoint
   ◆ Insert Tracepoint
   ⏱ Insert Temporary Breakpoint
   ⊡ Insert Dependent Breakpoint
```
37    ⊟class Person
```

Figure 3.27 – Inserting a dependent breakpoint

The **Insert Dependent Breakpoint** option will display the **Breakpoint Settings** window, where you will be asked which breakpoint you want to depend on to launch the dependent breakpoint. In our example, we will select the only breakpoint that is part of our project, as shown in the following screenshot:

Figure 3.28 – A list of breakpoints on which we can depend for dependent breakpoint execution

Once the **Breakpoint Settings** window is closed, you will see how a special breakpoint is created that (if you hover over it) will tell you on which other breakpoint it depends, as shown in the following screenshot:

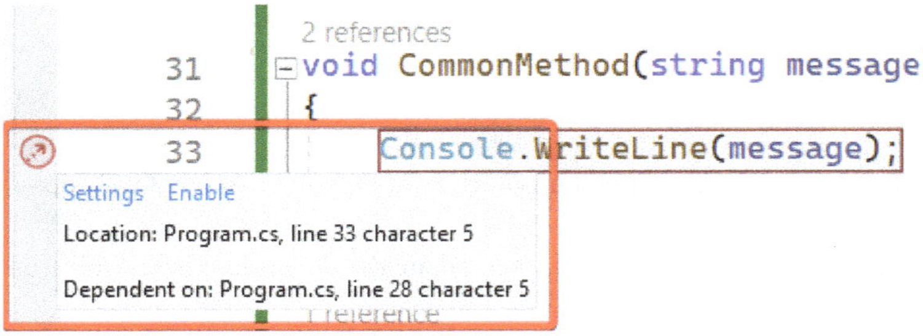

Figure 3.29 – A view of a dependent breakpoint

Finally, when running the application, you will see how it stops on the NewMethod call first, not CommonMethod, even though it was invoked first. If you continue the execution, the dependent breakpoint will stop, showing the **After invocation of NewMethod**() message, as shown in *Figure 3.30*:

Figure 3.30 – The execution of a dependent breakpoint

As you can see, a dependent breakpoint can be very helpful to avoid constant method invocation.

Now that you know about dependent breakpoints, let's take a look at another new breakpoint type called temporary breakpoints.

Note

Because a dependent breakpoint must be configured from the **Breakpoints** window, there is no shortcut associated with it. However, it is possible to configure it manually from the breakpoint configuration, as shown in *Figure 3.28*, by pressing the *Alt + F9*, then *C*, keys immediately after placing a breakpoint without any configuration.

Temporary breakpoints

Temporary breakpoints, as their name indicates, are breakpoints that are automatically deleted once they are executed. To place one of these breakpoints, you must right-click on the breakpoint line and select the **Insert Temporary Breakpoint** option, as shown in *Figure 3.31*:

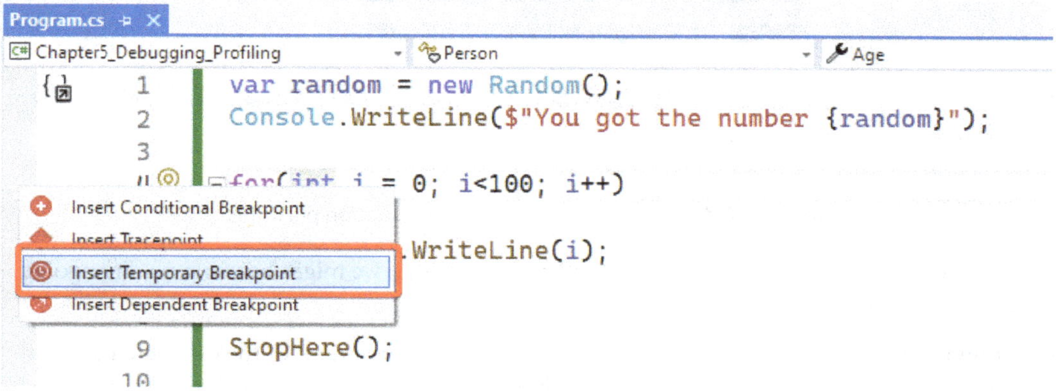

Figure 3.31 – The menu to insert a temporary breakpoint

If you run the application, you will see that once the application has stopped at the position where the temporary breakpoint was placed, that breakpoint will be automatically deleted. A temporary breakpoint can be used when we want to evaluate, the first iteration of a cycle, for example.

> **Shortcut**
>
> To insert a temporary breakpoint quickly, use the *F9 + Shift + Alt* keys, followed by the *T* key.

Now that we have examined the different types of breakpoints available in VS, let's see how we can take advantage of them using different inspection tools.

Breakpoint groups

Breakpoint groups are an excellent option for grouping breakpoints and enabling or disabling them according to the situation. Suppose you have spent time placing breakpoints in a general way in the application, but in a new session, you want to test a specific part of the application without going through all the general breakpoints. In such cases, group breakpoints are of great help.

Let's look at this practically. In the practice project, go to the `BreakpointGroupsDemo.cs` file; this is a class that calculates the age from your day, month, and year of birth. Here, I have placed breakpoints after prompting the user for their day, month, and year of birth, to debug this information, as shown in *Figure 3.32*:

```
 3    public class BreakpointGroupsDemo
 4    {
          1 reference
 5        public void Run()
 6        {
 7            Console.WriteLine("Write the day of your birth");
 8            var day = Console.ReadLine();
 9            Console.WriteLine("Write the month of your birth");
10            var month = Console.ReadLine();
11            Console.WriteLine("Write the year of your birth");
12            var year = Console.ReadLine();
13
14            var birthDate = new DateTime(int.Parse(year), int.Parse(month), int.Parse(day));
15
16            CalculateAge(birthDate);
17        }
18
```

Figure 3.32 – Reference breakpoints for demonstration purposes

This breakpoint placement simulates the set of breakpoints that we might be using on an ongoing basis in an application.

Now, suppose that at some point, we encounter an error in the printout of the calculated age and we want to debug the `CalculateAge` method but without removing the breakpoints we marked previously. To accomplish this, I will create the necessary breakpoints in the `CalculateAge` method, after which I will navigate to **Debug** | **New Breakpoint** | **Breakpoint group** or the **Breakpoints** window and add a group via **New** | **Breakpoint group**:

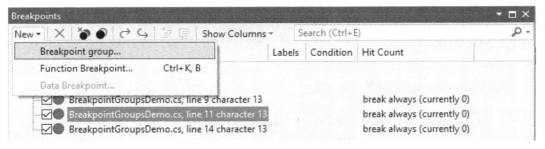

Figure 3.33 – Adding a breakpoint group

Either of the preceding options will prompt for the name of a new breakpoint group; in my case, I am going to call it `CalculateAgeGroup`, as shown in *Figure 3.34*:

New Breakpoint Group   ×

Name: CalculateAgeGroup

☐ Only enable when the following breakpoint is hit:

                                                    OK   Cancel

Figure 3.34 – Window for naming a breakpoint group

Next, we must run the application. This will bring up the **Breakpoints** window, where you will see the new group that was created. Here, simply drag the breakpoints you are interested in into the group that was created, as shown in *Figure 3.35*:

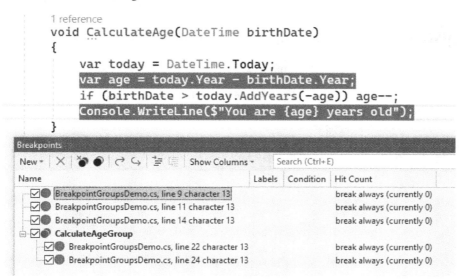

Figure 3.35 – The window with the new breakpoint group created

The advantage of grouping breakpoints is that we can check and uncheck the group box to enable or disable breakpoints when we need them. We could even create a second group to group the rest of the breakpoints and have better control, as shown in *Figure 3.36*:

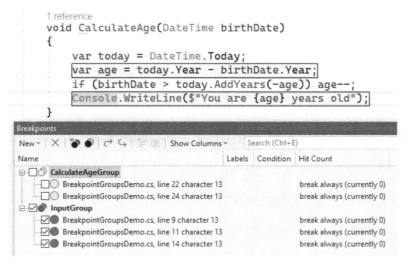

Figure 3.36 – Grouping breakpoints into different breakpoint groups

In this section, you learned about the different types of breakpoints available in VS so that you can perform quick debugging by knowing how to choose the right type for the situation you are in.

Now, let's look at some tools that VS offers to inspect information while in a debugging state.

Quickly debugging with Run to click and Force to click

One last feature I would like to talk about in this section is the **Run to click** and **Force to click** options.

These options are very useful to get to the previous execution of a specific line of code. To activate them, you just need to be in a debugging session and position your cursor next to the line you want to run to. A green glyph will appear:

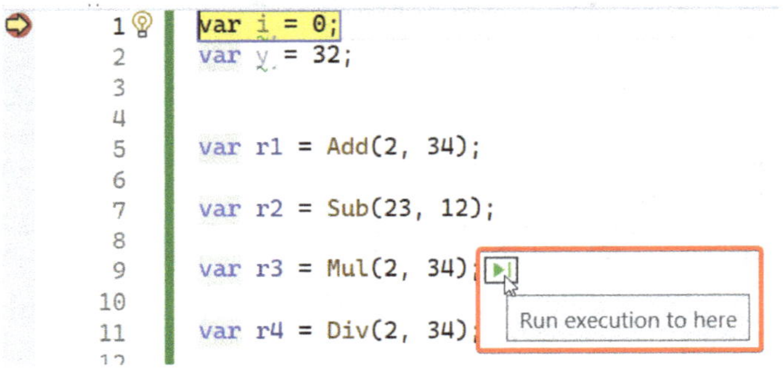

Figure 3.37 – The Run execution to here feature

Pressing this green glyph will activate the **Run to click** feature, which will cause the code to be executed until it reaches the desired line. Note that if there are breakpoints along the path to the line to be executed, the execution will stop at each of them.

On the other hand, if you press the *Shift* key, the icon will change to a double green glyph, indicating that you will use the **Force to click** feature. The difference is that in **Force to click** mode, all breakpoints will be ignored, and you will arrive just before the line you have selected is executed.

Inspection tools for debugging

When working with breakpoints, we must know where to find the information we want to visualize and whether that information is correct or not. For this, within VS, we have a series of windows that will allow us to visualize different types of information. So, let's take a look at the most important ones.

The Watch window

The **Watch** window allows us to keep track of the values of variables or properties while we execute our code step by step. This window is especially useful when we have pieces of code that are repeated several times, such as cycles or common methods. To access this window, we must place a breakpoint in the code and execute the application.

Once the application stops at the breakpoint, we will be able to deploy the **Debug | Windows** menu. This will show us a set of new debugging windows that we can only access while running the application. Let's select the **Watch** option so that we can choose a window, as shown in *Figure 3.38*:

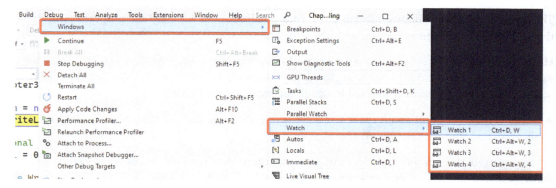

Figure 3.38 – The menu to reach the Watch option

Once the window has been opened, we can add the names of different variables and properties that we want to monitor. So long as we have entered a valid variable name for the scope we are in, we will be shown its corresponding value in the **Value** column, as shown in *Figure 3.39*:

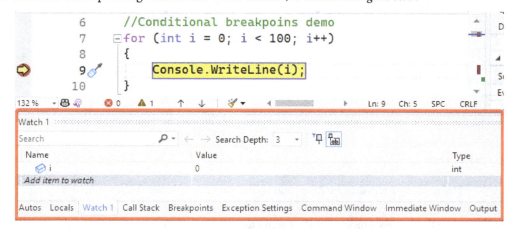

Figure 3.39 – Monitoring variable values and properties through the Watch window

Another very simple way to add variables to a **Watch** window is to right-click on the variable you want to monitor when the application is running and select the **Add Watch** option.

> **Shortcut**
> To access any **Watch** window quickly, you can use the *Ctrl + Alt + W* shortcut, followed by the window number (from *1* to *4*).

The Autos and Locals windows

The **Autos** and **Locals** windows allow us to view information about variables and properties without the need to add them somewhere, as with the **Watch** window.

However, these windows have a specific scope. While the **Autos** window shows the value of the variables around the breakpoint we have placed, the **Locals** window only shows values for the current scope – that is, the function or method in which the breakpoint is located.

Another important point about these windows is that they will only be shown when we run the application and after we have placed a breakpoint.

To demonstrate this window pair, let's put a breakpoint in the `for` loop, as shown in *Figure 3.39*. Once we invoke the method and only when the application is running, we will be able to display these windows through the **Debug** | **Windows** | **Autos** menu and the **Debug** | **Windows** | **Locals** menu:

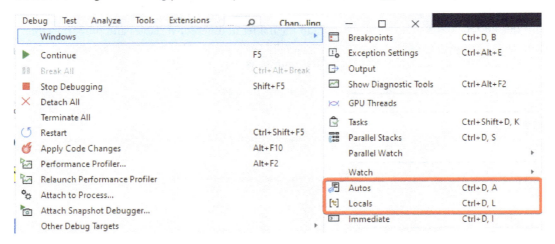

Figure 3.40 – Accessing the Autos and Locals windows from the Debug menu

First, let's examine the **Locals** window. As shown in *Figure 3.41*, the window contains all the available local variables, including the `i`, `random`, `person`, and other variables:

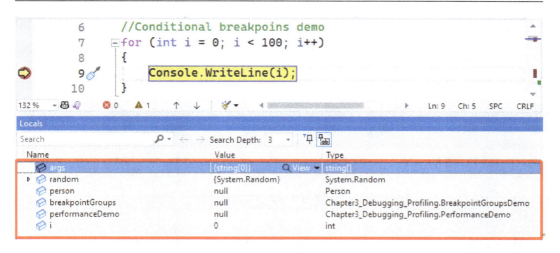

Figure 3.41 – The Locals window

On the other hand, if we look at the **Autos** window, we will only see the content of the i variable since it is the only information that is in the scope of the breakpoint. This is the for loop:

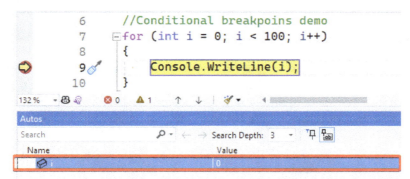

Figure 3.42 – The Autos window

This pair of windows is tremendously useful for viewing all the available information about our variables at a glance.

Shortcut

It is possible to display these windows through the following shortcuts:

Autos: *Ctrl + D*, followed by the *A* key

Locals: *Ctrl + D*, followed by the *L* key

These are some of the most important tools that can help you obtain information on the status of your applications. Now, let's see how VS can help us detect low performance or bottlenecks in our applications.

Measuring app performance with profiling tools

Application bottlenecks are a problem you will encounter during your career as a developer on any platform. It is for this reason that you must learn to manage the IDE tools that allow you to analyze the performance of your applications and objects created during the life cycle of your application. In this section, you will learn about these tools through a case study.

Analyzing the performance of an application

The first tool we will talk about is the **CPU Usage** diagnostic tool, which allows us to measure CPU usage during the execution of an application. This tool works thanks to the different breakpoints and debugger stops. Let's see this practically.

In the project you downloaded from GitHub, in the `Program.cs` file, you will find the following code section:

```
var performanceDemo = new PerformanceDemo();
performanceDemo.Run();
Console.WriteLine("Press any key to exit...");
```

This example simulates the invocation of a map service, which gets information from a thousand coffee shops and places an icon referring to the coffee shop on a map. Of course, we will use the console to simulate drawing the coordinates so that we don't overcomplicate this example. All the logic can be found in the `PerformanceDemo.cs` file, but here's the code for execution:

```
public void Run()
{
    MapService mapsService = new MapService();
    var businesses = mapsService.GetBusinesses();
    DrawBusiness(businesses);
}
```

In the preceding code, you can see that a `MapService` instance is created, which obtains a set of businesses and, finally, performs the drawing simulation through `DrawBusiness`.

Practically, let's suppose that we have detected that the application, when invoking the map service, takes much longer than expected to obtain information about the coffee shops, so we want to detect the method that is causing the problem.

Let's open the `MapService.cs` file and find the `GenerateBusinesses` method, which contains the logic to obtain the name, coordinates, business type, and business type icon. To activate the CPU usage tool, we must place two breakpoints – one indicating where we will start the analysis and another indicating where the analysis will end. In our example, I will place the first breakpoint at the beginning of the `GenerateBusinesses` method and the second one at the return of `businesses`, as shown in *Figure 3.43*:

```
     1 reference
17   private List<Business> GenerateBusinesses()
18   {
19       var businesses = new List<Business>();
20       Random random = new Random();
21
22       for (int i = 0; i < 1000; i++)
23       {
24           var business = new Business()
25           {
26               Name = GetBusinessName(i),
27               XPosition = GetRandomPosition(),
28               YPosition = GetRandomPosition(),
29               BusinessType = BusinessType.CoffeeShop,
30               Icon = GetIcon()
31           };
32           businesses.Add(business);
33       }
34       return businesses;
35   }
```

Figure 3.43 – Placing breakpoints in the GenerateBusinesses method

Once we have placed these breakpoints, the next step is to start running the application in debug mode. This will open the **Diagnostic Tools** window; if you do not see this window, you can open it by going to **Debug | Windows | Show Diagnostics tools**. In this window, you must go to the **CPU Usage** tab and activate the **Record CPU Profile** option, as shown in *Figure 3.44*:

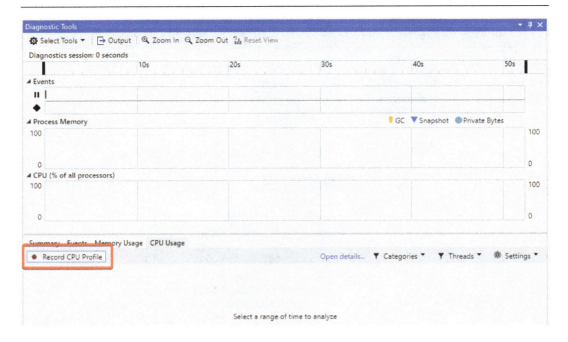

Figure 3.44 – Enabling CPU profile recording

This will start the process of recording the CPU profile until another breakpoint is reached; in our example, we will click on **Continue** to reach the second breakpoint. You must be patient as the process takes about 30 seconds to finish.

Once the second breakpoint has been reached, an analysis will be created where you can see the name of the main functions that are performing the most workload, as well as information such as the total percentage of CPU time used by the function, as shown in *Figure 3.45*:

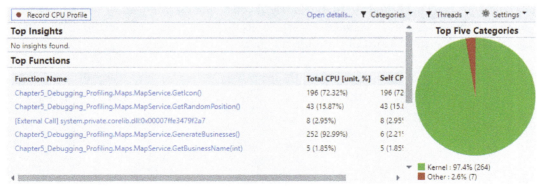

Figure 3.45 – Analysis of CPU utilization by function

In our example, we can see that the functions have been sorted according to CPU consumption – that is, **GetIcon**, **GetRandomPosition**, **GenerateBusiness**, and **GetBusinessName** – which can give us a great idea of which methods to optimize.

Now, let's look at the second tool that can help you find memory consumption problems.

Analyzing memory usage in your apps

Another tool that VS has to improve the performance of applications is the **Memory Usage** diagnostic tool. This tool can help you identify when you are saturating the memory with unnecessary objects, such as when you load files and they are not released.

To use this tool, it is necessary to take a snapshot between two points to know how many objects have been created during the execution of the code. To see this practically, let's go back to the MapService. cs file and reposition the breakpoints, as shown in *Figure 3.43*.

Next, run the application; it should stop at the first breakpoint. Now, in the **Diagnostic tools** window, make sure that the **Memory Usage** option is selected, as shown in *Figure 3.46*:

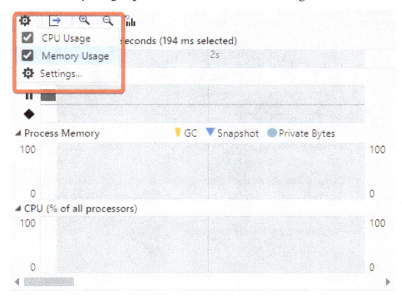

Figure 3.46 – Activating the Memory Usage option

Once you've done this, go to the **Memory usage** tab and press the **Take Snapshot** button, as shown in *Figure 3.47*. In the same figure, you can see how information has been added about the objects that were created at this point in the application, as well as the size of the heap. Once the snapshot has been taken, press the **Continue** button to reach the second breakpoint:

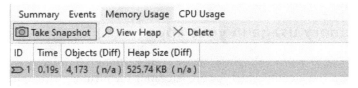

Figure 3.47 – Taking a snapshot of memory usage

When the second breakpoint is reached, go back to the **Memory Usage** tab and take a new snapshot; this will add new information to the object table, as shown in *Figure 3.48*. In our example, you can see that the number of objects and the size of the heap have increased considerably:

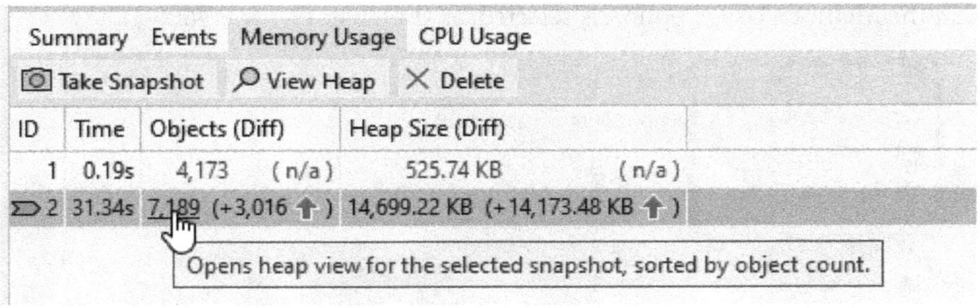

Summary	Events	Memory Usage	CPU Usage		
ID	Time	Objects (Diff)		Heap Size (Diff)	
1	0.19s	4,173	(n/a)	525.74 KB	(n/a)
2	31.34s	7,189	(+3,016 ↑)	14,699.22 KB	(+14,173.48 KB ↑)

Opens heap view for the selected snapshot, sorted by object count.

Figure 3.48 – Selecting the option to open the detail window of the number of objects used

If you want to see more information about the existing objects, you can click on one of the numbers, as shown in *Figure 3.49*. This will open a new tab in VS that shows the data of the objects in memory at the point of the snapshot; for example, we can see that there are 1,000 instances of the **Business** object and that they are using about 14 MB of memory:

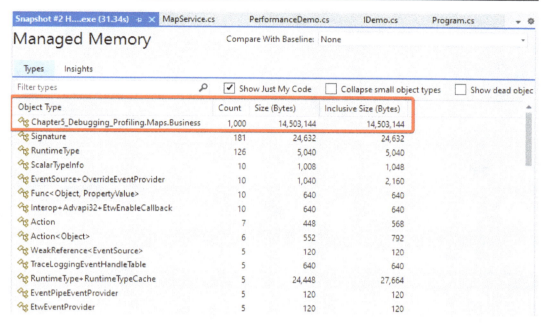

Figure 3.49 – The Managed Memory window

Without a doubt, these CPU and memory usage measurement tools are fabulous for detecting any performance problems in any application developed with VS.

Summary

VS has many options for debugging source code. In this chapter, we learned what breakpoints are, the different types of breakpoints, and the associated windows that we can activate to keep track of data in variables and properties. This information is of utmost importance since it will help you solve bugs in your code when you face them.

Likewise, you learned how to measure the performance of your applications, both in terms of CPU usage to detect functions that have the highest workload, as well as to detect places in the code that may be creating object references that are not released.

In the next chapter, you will learn about the concept of code snippets, which allow you to reuse common pieces of code across different projects and quickly adjust them to suit your needs.

Part 2:
Tools and Productivity

In this part, you will learn how to take advantage of the tools provided by Visual Studio 2022 and how to improve your productivity with some tips and hacks.

This part has the following chapters:

- *Chapter 4, Adding Code Snippets*
- *Chapter 5, Coding Efficiently with AI and Code Views*
- *Chapter 6, Using Tools for Frontend and Backend Development*
- *Chapter 7, Styling and Cleanup Tools*
- *Chapter 8, Publishing Projects*

4

Adding Code Snippets

Coding is a lovely activity, but at times, some repeated statements are used to solve a few already known situations, which makes coding a more monotonous process. **Code snippets** represent a good resource for reusing pieces of code where desirable. By default, VS has some code snippets that we can use while we are coding, although it is also possible to manage them within VS, either to add new code snippets or to remove them.

In this chapter, we will learn about how VS helps us to write code faster by using code snippets and how to create our own versions.

We will review the following topics and functionalities for code snippets:

- What are code snippets?
- Creating code snippets
- Deleting code snippets
- Importing code snippets

Let's start recognizing the concept of code snippets and how they work in VS.

Technical requirements

In order to complete the demos in this book chapter, you must have previously installed VS 2022 along with the web development workload, as shown in *Chapter 1, Getting Started with Visual Studio 2022*. You can download the project used in this chapter at the following link: `https://github.com/PacktPublishing/Hands-On-Visual-Studio-2022-Second-Edition/tree/main/Chapter%204`.

What are code snippets?

Code snippets are a simple and easy way to reuse code by creating templates that generate common statements, such as conditionals, loops, or comment structures.

VS has many code snippets by default for almost all the supported technologies and programming languages.

> **Important note**
>
> Code snippets are a common concept in software development. Almost all IDEs and code editors provide code snippets or have extensions to include code snippets.

In order for you to follow the different case studies used throughout this chapter, you can create a new project by using the ASP.NET Core with React.js template or downloading the resources mentioned in the *Technical requirements* section, where you will find a folder called Start with the initial files of the project.

Once you have the project open in VS, you will need to create a new condition to return an empty collection using the Get method by navigating to the WeatherForecastController.cs file. Just write the word if at the beginning of the Get method to see the code snippet suggested by VS (see *Figure 4.1*):

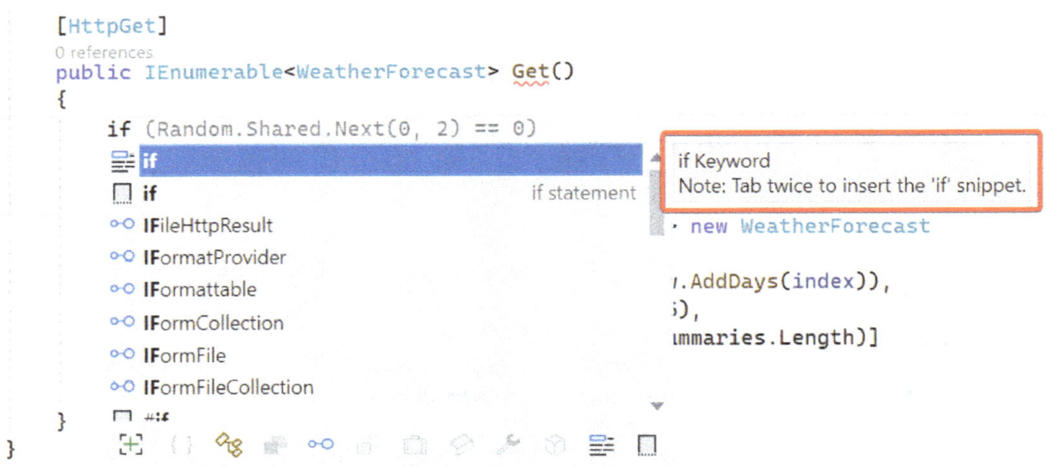

Figure 4.1 – The code snippet for the "if" statement suggested by VS

Since a conditional statement is a very common piece of code, VS gives you the option to create this code quickly. You can click on if or continue writing if you don't want to perform any action. You can also press *Tab* twice to create the if statement automatically.

There is an option in the intelligent code completion (also called **IntelliSense**, which we will talk about more in *Chapter 5, Coding Efficiently with AI and Code Views*) suggestions, where you can see all the code snippets filtered by the characters that you wrote. See the code snippets filter marked in red in *Figure 4.2*:

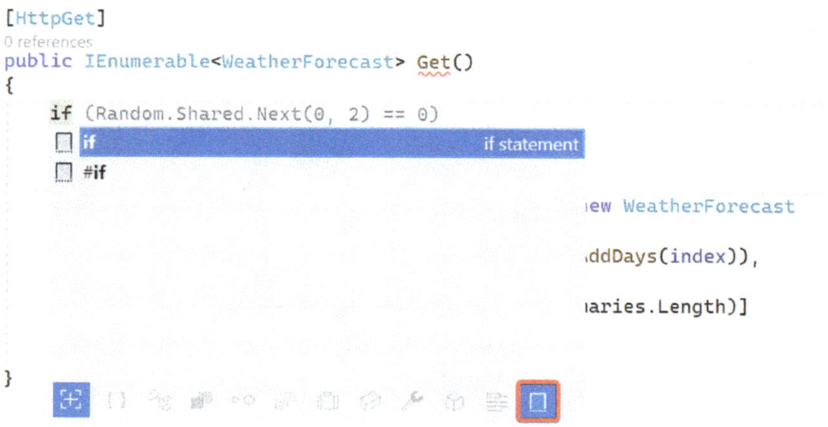

Figure 4.2 – The code snippets filter in VS

When the code snippet for the `if` condition is highlighted, you can press the *Tab* key twice to generate the code of the `if` statement, including the brackets (see *Figure 4.3*):

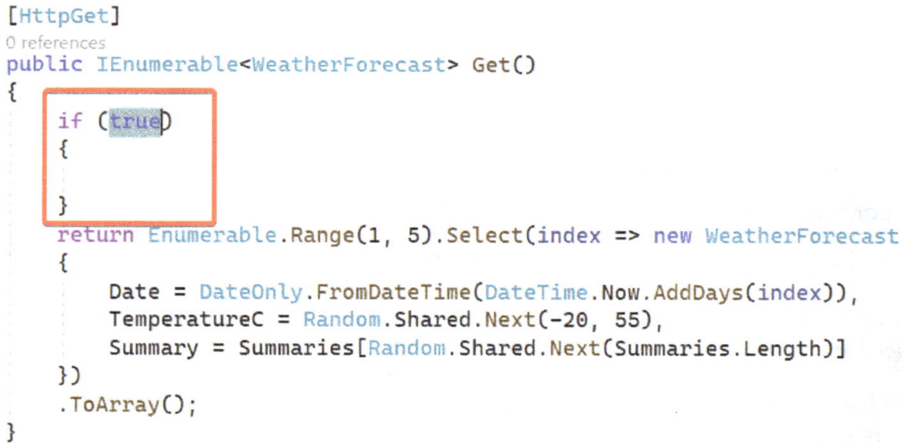

Figure 4.3 – The "if" statement created by VS

You will get the `if` statement, including the braces and `true` as a default value. You need to replace `true` with your condition. In this case, you can add a condition to return an empty collection when the operating system is Linux:

```
[HttpGet]
public IEnumerable<WeatherForecast> Get()
{
```

```
if (OperatingSystem.IsLinux())
{
    return new List<WeatherForecast>();
}

return Enumerable.Range(1, 5).Select(index =>
    new WeatherForecast
    {
        Date = DateTime.Now.AddDays(index),
        TemperatureC = Random.Shared.Next(-20, 55),
        Summary = Summaries[Random.Shared.Next
            (Summaries.Length)]
    })
    .ToArray();
}
```

In the preceding code block, we added a condition in the Get method before the default logic to check whether the operating system on which the app is running is Linux or not. Within the condition, we return an empty list.

There are many useful code snippets included as default for C#, but the following are the most popular:

- try: Creates a structure for a try/catch statement
- for: Generates a for statement using the local i variable
- ctor: Creates the constructor of the class automatically
- switch: Generates a switch statement
- prop: Creates a new property in the current class

You can try some of these code snippets in C# code to see the code generated by VS and use them when the need arises.

Let's see another example using a CSS file. Navigate to the ClientApp | src | custom.css CSS file (see *Figure 4.4*):

```
custom.css  ⊹ ×  Weather...oller.cs                        ⚙  Solution Explorer              ▼ ♁ ×
    1      /* Provide sufficient contrast against ⊹      ⌨ ⌕ · ↰ ▤ ⌸ ⌸ · ⚒ ▱
    2     ⊟a {                                                  Search Solution Explorer (Ctrl+`)        ₽ ·
    3         color: #0366d6;                                        JS AppRoutes.js
    4     }                                                          📄 custom.css
    5                                                                JS index.js
    6     ⊟.btn:focus, .btn:active:focus, .btn-li                    JS reportWebVitals.js
    7         box-shadow: 0 0 0 0.1rem white, 0 0                    JS service-worker.is
    8     }                                                   Solution Explorer  Git Changes
    9
   10     ⊟code {                                            Properties                       ▼ ♁ ×
   11         color: #E01A76;
   12     }                                                   ▦ ▣↓ ⚒
   13
   14     ⊟.btn-primary {
   15         color: #fff;
   16         background-color: #1b6ec2;
   17         border-color: #1861ac;
   18     }
   19
146 %  · 🖳   ⊘ No issues found      ◄ ▭▭   ►   Ln: 1  Ch: 1  SPC  CRLF

Error List  Output
```

Figure 4.4 – The custom.css file loaded in VS

In the `custom.css` file, you can write the `columns` property and see what VS suggests as a code snippet for it. See the code snippets suggested for this demo in *Figure 4.5*:

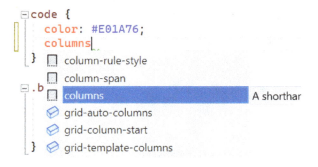

Figure 4.5 – The code snippets suggested by VS when you type the word "columns"

Once again, by pressing the key *tab* twice, you can generate the code for the `columns` property automatically. See the code generated in *Figure 4.6*:

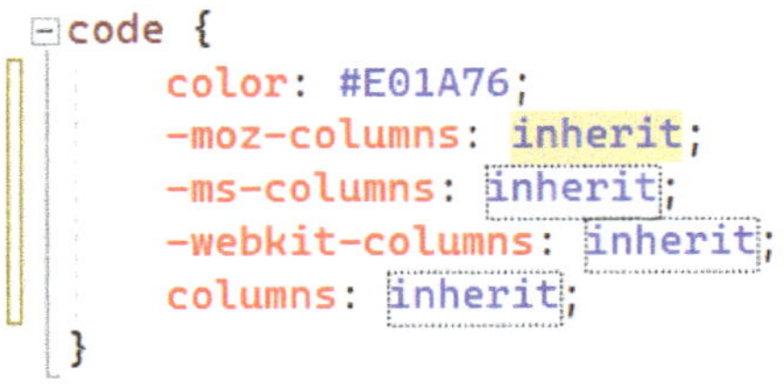

```
code {
    color: #E01A76;
    -moz-columns: inherit;
    -ms-columns: inherit;
    -webkit-columns: inherit;
    columns: inherit;
}
```

Figure 4.6 – The code generated by the code snippet

In this case, VS is going to generate four properties, one per browser, to ensure the code is compatible with different browsers, including old and new versions (Chrome, Mozilla, and others). We can keep the default `inherit` value because the proposal of inserting the `columns` code snippet is used just to see how VS makes the code.

You are now ready to use code snippets in VS. You can identify which suggestions by VS are code snippets and how to filter them. Now, it's time to learn how to create your own code snippets and use them in your code.

Creating code snippets

In order to create a code snippet in VS, we need to create a file with the `snippet` extension. This file has an XML format, and there is a base template that we can update to include the information for our code snippet. The following code is a template example:

```xml
<?xml version="1.0" encoding="utf-8"?>
<CodeSnippets xmlns="http://schemas.microsoft.com/
    VisualStudio/2005/CodeSnippet">
    <CodeSnippet Format="1.0.0">
        <Header>
            <Title></Title>
            <Author></Author>
            <Description></Description>
            <Shortcut></Shortcut>
        </Header>
        <Snippet>
            <Code Language="">
                <![CDATA[]]>
            </Code>
        </Snippet>
    </CodeSnippet>
</CodeSnippets>
```

Let's review all the properties in this XML and understand how to create our first code snippet.

In the `Header` section, we have the following:

- `Title`: Name or general information
- `Author`: Creator or author
- `Description`: What your code snippets do
- `Shortcut`: The shortcut to call the code snippet when you are typing

In `Snippet`, we have the following:

- `Language`: The programming language for the code, with values for `VB`, `CSharp`, `CPP`, `XAML`, `XML`, `JavaScript`, `TypeScript`, `SQL`, and `HTML` being possible
- `[CDATA[]`: Contains the source code in the specified language

Now, we can create a code snippet to detect whether the operating system where the code is running is Linux or not (you can find this code snippet in the repository marked in the *Technical requirements* section):

```xml
<?xml version="1.0" encoding="utf-8"?>
<CodeSnippets xmlns="http://schemas.microsoft.com/VisualStudio/2005/
CodeSnippet">
    <CodeSnippet Format="1.0.0">
        <Header>
            <Title>If Linux condition</Title>
            <Author>Myself</Author>
            <Description>Conditional to now if the
                operating system is Linux</Description>
            <Shortcut>ifln</Shortcut>
        </Header>
        <Snippet>
            <Code Language="CSharp">
                <![CDATA[if (OperatingSystem.IsLinux())
                {
                    return new List<WeatherForecast>();
                }]]>
            </Code>
        </Snippet>
    </CodeSnippet>
</CodeSnippets>
```

You can create a new folder in documents or any folder; in my case, I have created a folder called `CodeSnippets`, saving the document in it as `codesnippet.snippet` (see *Figure 4.7*):

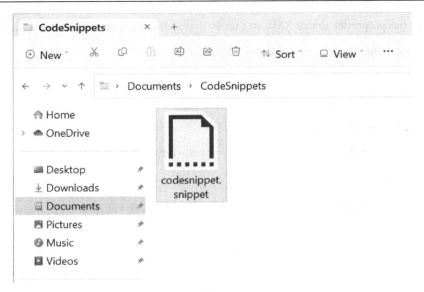

Figure 4.7 – The code snippet file in Windows explorer

Now, the last step is to add the CodeSnippets folder to the code snippet section in VS. Navigate to **Tools | Code Snippets Manager**, and in the **Language** dropdown, select **CSharp** (see *Figure 4.8*):

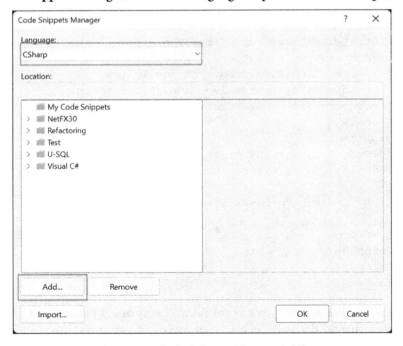

Figure 4.8 – Code Snippets Manager in VS

> **Important note**
> You can use the *Ctrl + K* shortcut followed by *Ctrl + B* to open the **Code Snippets Manager**.

The **Language** option includes all the programming languages and technologies supported by VS, depending on the workload installed.

You can click on **Add...** and select the folder where your code snippet was created.

After adding the folder, you will see a new folder in the list, including the new code snippet. If you select this code snippet, you will see the details on the right panel, as shown in *Figure 4.9*:

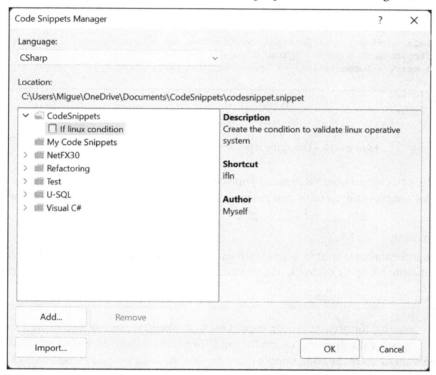

Figure 4.9 – The code snippet details for the if linux condition

Now, you are ready to use your code snippet in any C# file. Open the `WeatherForecastController.cs` file and delete the following code section:

```
if (OperatingSystem.IsLinux())
{
    return new List<WeatherForecast>();
}
```

Now, in the place where the deleted piece of code was, that is, within the `Get` method, type `ifln`, which is the shortcut for the assigned code snippet:

```
[HttpGet]
0 references
public IEnumerable<WeatherForecast> Get()
{
    ifln
    ☐ ifln                                                    If Linux condition
    ⬚  () ⚙ ☁ ∞ ☁ ☐ ⬚ ⚙ ☰ ☰ ☐         Conditional to now if the
                                                                 operating system is Linux
                                                             Note: Tab twice to insert the 'ifln' snippet.
    return Enumerable.Range(1, 5).Select(index => new WeatherForecast
    {
        Date = DateOnly.FromDateTime(DateTime.Now.AddDays(index)),
        TemperatureC = Random.Shared.Next(-20, 55),
        Summary = Summaries[Random.Shared.Next(Summaries.Length)]
    })
    .ToArray();
}
```

Figure 4.10 – Using the if linux condition code snippet in VS

In *Figure 4.10*, you can see how VS suggests your new code snippet, and in the tooltip, there is the description that you provided. As usual, you can press *Tab* twice to generate the code for the code snippet.

> **Important note**
>
> For more information about how to create and design code snippets, refer to the official documentation: https://docs.microsoft.com/visualstudio/ide/code-snippets.

Once we have seen that the preceding code snippet works correctly, we realize that we are a bit limited in terms of its use since it will always return the same list when we use it even if we are in a different project. If we want to make the code snippet more flexible, the ideal proposal would be to be able to quickly replace the object to be returned so that it can be a list of objects, a string, or a Boolean value.

In order to achieve this, we must add a section called `Declarations`. Here's an example:

```
<Snippet>
  <Declarations>
    <Literal Editable="true">
      <ID>object</ID>
      <ToolTip>The object to be returned</ToolTip>
      <Default>object</Default>
    </Literal>
```

```
  </Declarations>
  <Code Language="csharp" Delimiter="$"><![CDATA[if (OperatingSystem.
IsLinux())
          {
              return $object$;
          }]]></Code>
</Snippet>
```

The `Declarations` section allows us to indicate the elements in the code that can be replaced through the `ID` tag. For example, in the preceding code snippets, we have replaced `List<WeatherForecast>()` with `$object$` inside the `Code` section in order to indicate that this identifier can be replaced by the user.

We have also added the `ToolTip` tag to give information to the user about what the code section to be replaced will do and a `Default` tag, which will be the default code inserted before the replacement.

If we re-import the snippet and use it, you will see that this time, the cursor is positioned over the word `object`, which is the identifier that we have set to be replaced in the code, as you can see in the following figure:

```
if (OperatingSystem.IsLinux())
{
    return object;
}
```

Figure 4.11 – Showing fast code replacement thanks to the Declarations section

So far, you have created your first code snippet and know how to create others that meet your needs. But there are also other actions that you can perform with code snippets. So, let's see how to delete a code snippet in the next section.

Deleting code snippets

Due to human error, we can add code snippets that we don't need, or maybe we can select the wrong code snippet. For these scenarios, VS has an option to delete code snippets. In order to see this option, navigate to **Tools | Code Snippets Manager** and select the `CodeSnippets` folder. This folder contains the code snippet that you included in the *Creating code snippets* section. If you used a different name, select the correct folder for you. You can see the **Remove** button location in *Figure 4.12*:

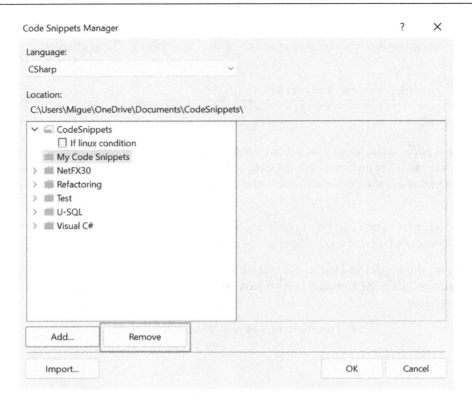

Figure 4.12 – The Remove button in Code Snippets Manager

The **Remove** button will delete the whole folder, including all the code snippets inside. In VS 2022, it's not possible to remove code snippets one by one, and therefore, we need to create a folder with a proper name for our code snippets. After removing the folder, VS will not suggest the code snippets anymore.

> **Important note**
> When you remove code snippets in VS, the original files and folders are not removed from your local system. Only the reference to the file in VS will be removed.

At this point, you know how to create and delete code snippets. We can also import code snippets in VS. So let's see how to do it.

Importing code snippets

If we want to include code snippets in a folder already created in the **Code Snippets Manager**, we can use the **Import…** option:

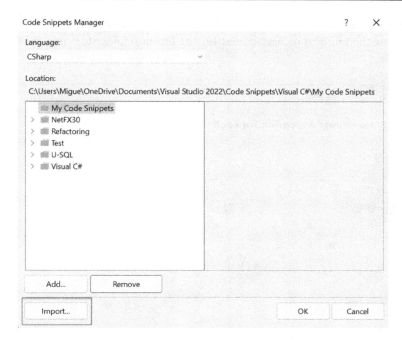

Figure 4.13 – The "Import…" button in Code Snippets Manager

After clicking on **Import…**, you need to select the code snippet that you want to import in the selected folder. There is a filter related to the .snippet extension in the modal (see *Figure 4.14*):

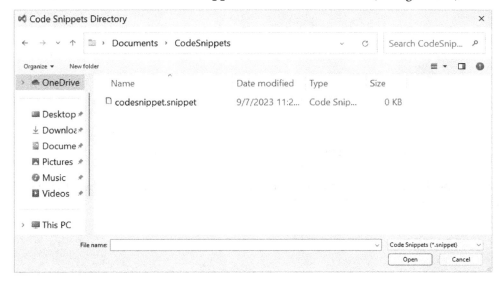

Figure 4.14 – Selecting a code snippet file (.snippet) in the filesystem

Select the code snippet created in the *Creating code snippets* section, and then click on **Open**. Finally, you must complete the import process by selecting the location folder for your code snippet and clicking on **Finish**:

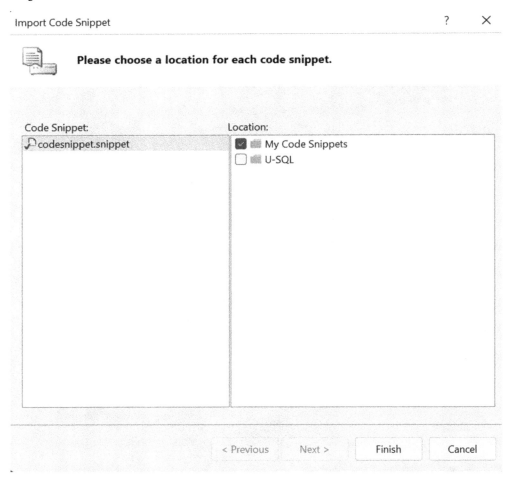

Figure 4.15 – The selection location for the imported code snippet

After completing the importation, you will see that the code snippet is added to the selected folder. See the imported code snippet in *Figure 4.16*:

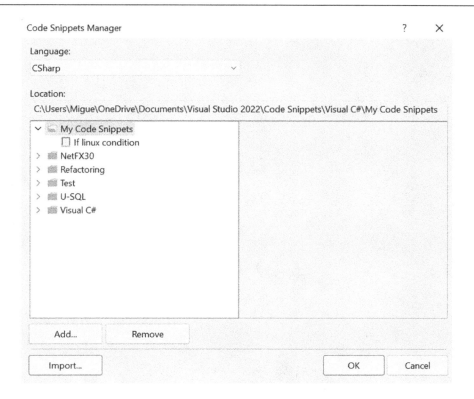

Figure 4.16 – The code snippet imported into the My Code Snippets folder

Importing code snippets is a great way to share our custom snippets with friends, colleagues, and coworkers. We can create amazing pieces of code for common scenarios and some special code closely related to our architecture or guidelines.

Summary

Now, you can use code snippets in VS and increase your productivity. You can identify which pieces of code are common in your architecture and use patterns to create your own code snippets to meet your requirements. Also, you know how to manage code snippets by using the functionalities to delete and import. After completing the demos in this chapter, you will recognize the importance of code snippets and why VS is a powerful IDE that helps developers write code faster.

In *Chapter 5, Coding Efficiently with AI and Code Views*, we will review the **artificial intelligence (AI)** tools included with VS and the ones we can add to the IDE, and we will learn how these tools can help us to write code faster and improve the syntax in some scenarios. You will also carry out some demos where AI will help you, allowing you to predict what action or statement you want to perform. So stay tuned!

5

Coding Efficiently with AI and Code Views

Artificial intelligence (**AI**) is a vast and interesting field that allows us to improve our lifestyle in one way or another. We see, hear, and use it every day, and if you don't believe it, ask yourself how many times you use the Google search engine throughout the day. Other places where you can find it are in photo-editing programs, where it is possible to remove, for example, the background of an image in an almost perfect way. Social networks are another perfect example of the use of artificial intelligence, as they are constantly processing the best recommendations for you to stay on them as long as possible.

Fortunately, AI has even reached new software development tools through predictive code integration, which allows us to choose the pieces of code we need at the right time. In VS, we have a powerful feature called **Visual Studio IntelliCode** that does this.

Similarly, with the recent launch of services such as **ChatGPT** and the close collaboration with Microsoft, a special model has been trained to help programmers perform their tasks faster called **GitHub Copilot**, which we will analyze in detail in this chapter.

Lastly, we have different visual tools and windows that can help us find relationships in our code and navigate through them efficiently.

In this chapter, we will cover the following main topics:

- Understanding CodeLens
- Working with code views
- Using Visual Studio IntelliCode
- Exploring GitHub Copilot
- Experiencing GitHub Copilot Chat

Technical requirements

To use IntelliCode, VS 2022 with the web development workload must be installed. You can find the code repositories at `https://github.com/PacktPublishing/Hands-On-Visual-Studio-2022-Second-Edition/tree/main/Chapter%205`.

Also, to be able to perform the procedure in the *Code maps* section, you will be required to use the Enterprise version of VS.

Likewise, the **Code Map** and **Live Dependency Validation** tools must be installed by the Visual Studio Installer. Select them as shown in *Figure 5.1*:

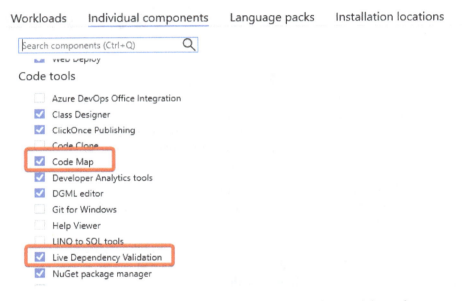

Figure 5.1 – Installation of the Code Map and Live Dependency Validation features

Now that we know the technical requirements, let's learn how to work with them to get the most out of VS.

Understanding CodeLens

CodeLens is a powerful set of tools that is useful for finding references in code, identifying relationships between your different components, seeing the history of changes in the code, related bugs, code reviews, and unit tests, and so on.

In this section, we will analyze the most important tools of this feature. Let's start by seeing how we can find references in our code.

> **Note**
>
> You can check if CodeLens is enabled by going to **Tools | Options | Text Editor | All Languages | CodeLens**, where you can see if the **Enable CodeLens** option is selected or not.

Finding references in code

CodeLens is presented in our code files from the first time we use VS. We can check this by going to any class, method, or property and verifying that a sentence appears, indicating the number of references in the project about it. In *Figure 5.2*, we can see that we have opened the WeatherForecastController.cs file, which shows us that three references have been found for the WeatherForecastController class:

```
[ApiController]
[Route("[controller]")]
3 references
public class WeatherForecastController
{
    private static readonly string[] Su
    {
```

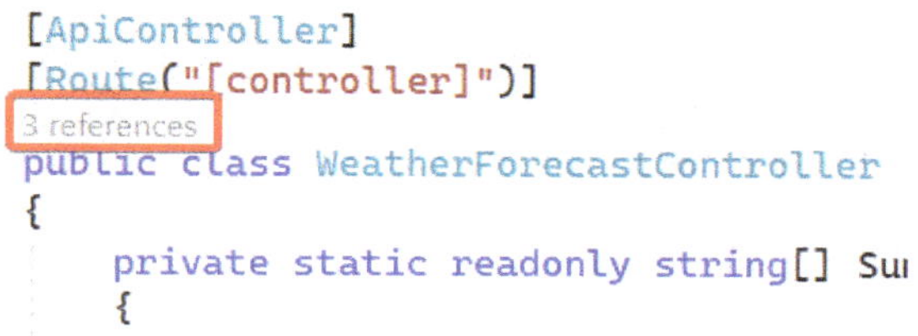

Figure 5.2 – References for the WeatherForecastController class

This means that the WeatherForecastController class is being used in three places in our project. If we proceed by clicking on the legend titled **3 references**, as shown in *Figure 5.2*, we will see all references that use this class. In *Figure 5.3*, continuing with the same example, we can see that it is used within the same class we are in—that is, the WeatherForecastController class—specifically in lines 14 and 16:

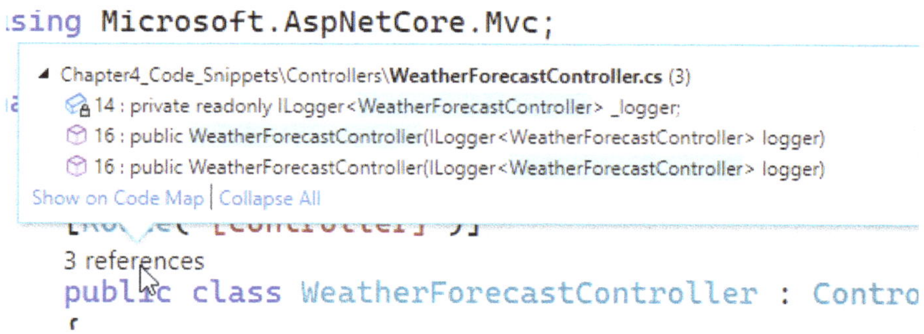

```
.sing Microsoft.AspNetCore.Mvc;

    ▲ Chapter4_Code_Snippets\Controllers\WeatherForecastController.cs (3)
        14 : private readonly ILogger<WeatherForecastController> _logger;
        16 : public WeatherForecastController(ILogger<WeatherForecastController> logger)
        16 : public WeatherForecastController(ILogger<WeatherForecastController> logger)
    Show on Code Map | Collapse All

3 references
public class WeatherForecastController : Contro
```

Figure 5.3 – The location in the code of the references in the WeatherForecastController class

In addition, we can also hover over on any of the lines found, which will show us a section of the four closest lines of code surrounding the reference. This way, we can get a better idea of the purpose of using the class, as shown in *Figure 5.4*:

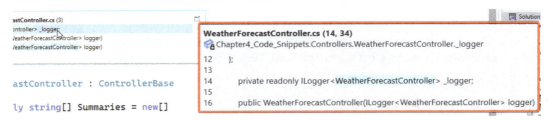

Figure 5.4 – A preview of a found code reference

This is quite useful if we are in a new project and need to quickly know what certain parts of the code do.

Now, let's see a utility belonging to CodeLens that will allow you to see relationships between code visually.

> **Important note**
> Sometimes, even if the number of references equals zero, there may be references to other GUI files, such as .xaml and .aspx files.

Code maps

Code maps are a way to visualize relationships in code in a fast and efficient way. This tool allows the creation of visual maps from the code, as its name indicates. With this tool, we will be able to see the structure of the entities, their different properties, and their relationships, which lets us know how much impact a change we make can have.

There are several ways to create code maps. The first one is by selecting the option **Architecture | New Code Map**. This will open a new document with a .dgml extension, where we will be instructed to drag files from the solution explorer, class view, or object browser, as shown in *Figure 5.5*:

Figure 5.5 – The empty code map file

Let's do a test—click on the **Class View** link to open the class window and then expand the SPAProject. Controllers namespace. This will show you the WeatherForecastController class, as shown in *Figure 5.6*:

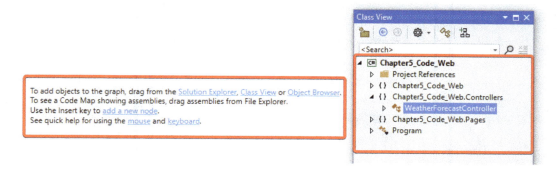

Figure 5.6 – Visualization of the WeatherForecastController class
about to be dragged into the code map file

Next, drag the WeatherForecastController class into the code map file. This will automatically generate a graph where we can see the dragged class, the namespace that contains it, and finally, the .dll in which it is hosted, as shown in *Figure 5.7*:

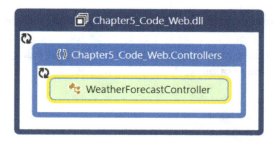

Figure 5.7 – The WeatherForecastController class in the code map

Additionally, if we expand the WeatherForecastController class in the diagram, we will see the members that are part of the class, such as its attributes and behavior, as well as the relationships that can be found as part of the same class:

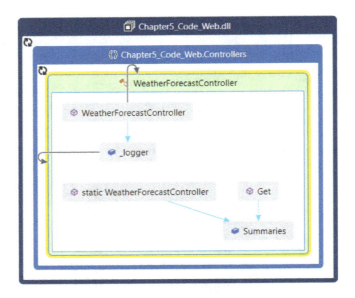

Figure 5.8 – The relationships found through Code Map

In *Figure 5.8*, we can see in action a code map of the `WeatherForecastController` class with all its members expanded. This shows us quickly how the fields, properties, and methods are related.

Another way to create a map code from the source code is to go to the file where the member we are interested in is located, such as the `WeatherForecast.cs` file. Once we have opened the file, we can position the cursor on a class, method, property, or field and right-click, which will show us the **Code Map | Show on Code Map** option, as shown in *Figure 5.9*:

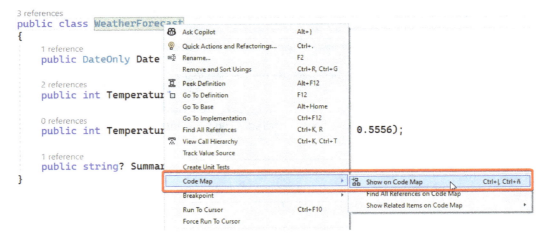

Figure 5.9 – The option to add a class to a code map from the context menu of a class

This option will create a new .dgml file or, if you have already created one, as in our case, add the reference with its respective relations in the previously opened file.

> **Important note**
>
> If you want to center the code map diagram, you can click on any empty area of the diagram to center it at any time. Likewise, if you double-click on any of the entities or members of the diagram, the corresponding code will open to view it next to the diagram.

As a result of adding the new class to the diagram, we can see that the WeatherForecast class is being used in the Get method of the WeatherForecastController class, as shown in *Figure 5.10*. This way, we have discovered the following relationship very easily:

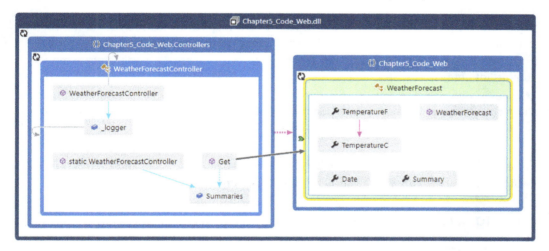

Figure 5.10 – The relationships between classes by Code Map

> **Important note**
>
> The arrows indicating relationships between entities in the code map diagram appear and disappear as entities are selected, allowing more space and understanding of the diagram. I encourage you to select each of the elements in the diagram so that you can see the complete relationship.

Finally, if we want to be able to see the relationships in our solution without having to add entity by entity, from the **Architecture** menu, we can select the **Generate Code Map for Solution** option, as we can see in *Figure 5.11*:

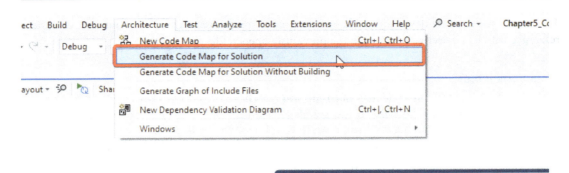

Figure 5.11 – The option in the menu to generate a code map at a solution level

This will start the process of generating the respective code map for the entire solution. Depending on the number of references in your code, the process may take more or less time.

> **Important note**
>
> Although code maps can only be created in VS Enterprise edition, it is possible to view them from any version of VS, including the Community version, but it is not possible to edit them from any version except the Enterprise Edition.

Now that we have seen how CodeLens can help us understand our code better and faster, let's look at the windows available in VS, which will allow us to work with our code easily.

Working with code views

In addition to CodeLens, there are several windows that can help us examine the classes of a project and its members in a quicker way. In this section, you are going to learn about them and how they can help you in breaking down the code of a project in VS.

Class View

Class View is a window that allows you to see the elements of a VS project, such as namespaces, types, interfaces, enumerations, and classes, allowing you to access each of these elements quickly. Perhaps, if you have worked with small projects in VS, you might not see the **Class View** window as being of much use because you can easily navigate between a few classes from the **Solution Explorer**. But like me, if you work with solutions that can have up to 20 projects or more with hundreds of classes throughout the projects, then the **Class View** is an excellent option for examining code.

To access this window, you must select the **View | Class View** option from the menu, which will display the **Class View** window, showing all the elements of the solution that is currently open, as shown in *Figure 5.12*:

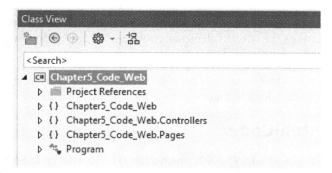

Figure 5.12 – The Class View window with a loaded project

As you can see, we can get a very quick idea of the structure of our project by seeing at a glance the namespaces into which the project has been divided. If we expand the nodes of each of the namespaces, we can also see the different classes that are part of these namespaces, as shown in *Figure 5.13*:

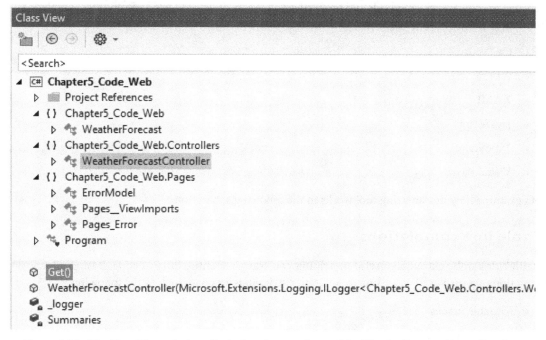

Figure 5.13 – The Class View window displaying the members of the WeatherForecastController class

In addition, if we select any element of our project, such as a class, we will be able to see the properties and methods that compose it in the lower part of the window (as shown in *Figure 5.13*).

Another great advantage of this window is that you don't even have to recompile the project to see the changes, as they will be made automatically and instantly.

At the top of the window in *Figure 5.13*, we can also see a series of buttons that we can use to create new folders, navigate between the selected elements, configure the display options, and possibly add a class to a code map file.

It is certainly an excellent window to navigate between the classes of our project. Now, let's look at a new feature called *IntelliCode*, which will allow us to write code more efficiently.

Using Visual IntelliCode

IntelliCode is the tool integrated into VS 2022, which allows you to write code faster, thanks to AI. It is a tool that has been trained with thousands of popular open source projects hosted on GitHub, and although it was already beginning to show a little of its potential in VS 2019, it is in VS 2022, where all the features have been implemented.

IntelliCode can suggest patterns and styles while you write code, giving you accurate suggestions according to the context in which you find yourself, so you can complete lines of code. IntelliCode is also able to show you the methods and properties you are most likely to use and supports completion in multiple programming languages, such as the following:

- C#
- C++
- XAML
- JavaScript
- TypeScript
- Visual Basic

Let's examine how this amazing tool works in the following subsections.

Whole line completions

IntelliCode can be extremely useful in helping you complete entire lines of code. Best of all, code predictions are displayed according to different entries in your code, such as the following:

- Variable names
- Function names
- IntelliSense options used
- Libraries used in the project

There are two ways to receive entire line completion hints in VS 2022:

- The first one is given automatically while you are writing code. In *Figure 5.14*, we can see this in action, as we start writing a new property in the `WeatherForecast.cs` file of type `string`:

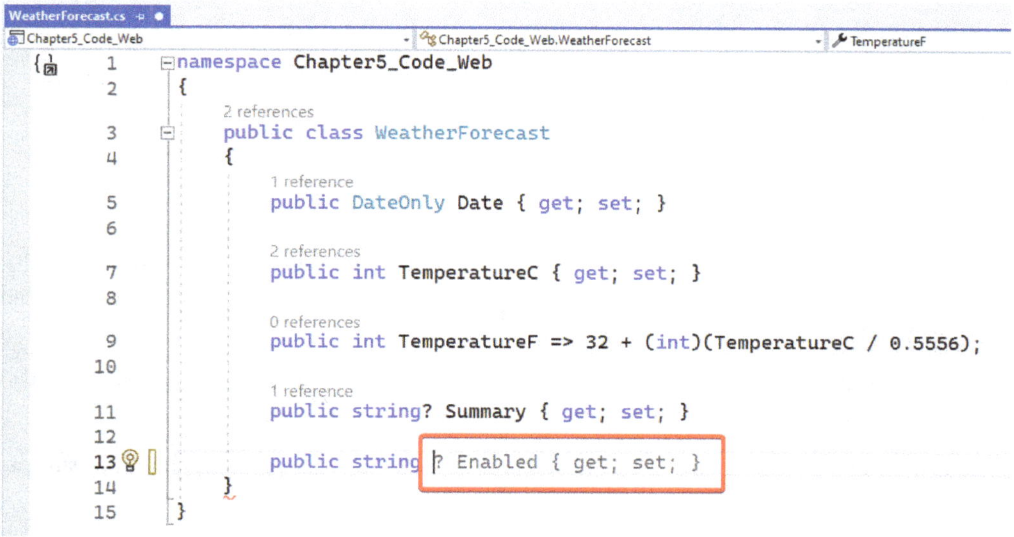

Figure 5.14 – IntelliCode suggesting a full line completion

According to everything learned by the IntelliCode model, it suggests a new property called `Description`, which we can accept by pressing the *Tab* key or reject by continuing to write code.

- The second method of line completion through the use of IntelliCode is done by selecting an item from the IntelliSense suggestion list. For example, if we create a constructor for the `WeatherForecast`, filter the properties, and type the letter S, a list of IntelliSense suggestions will appear. We can scroll through each of them, and in most cases, IntelliCode will show us auto-completion suggestions, as shown in *Figure 5.15*:

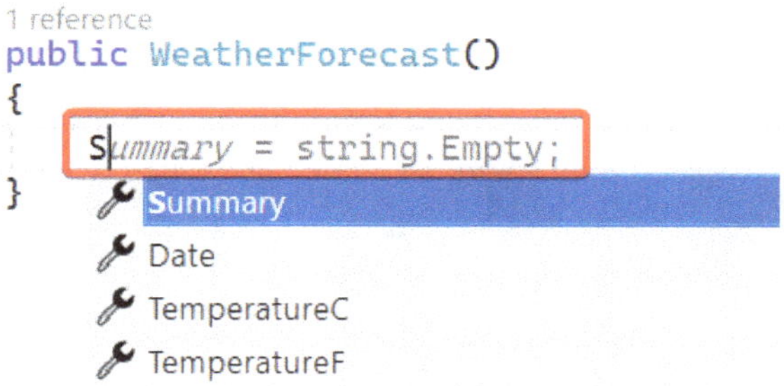

Figure 5.15 – IntelliCode recommending the completion of an IntelliSense element

We can accept the line by pressing the *Tab* key twice or continue writing code to ignore the suggestion.

Now that we have seen the two methods for line completion, let's see how IntelliCode can help us write code based on its suggestions.o

> **Note**
>
> In case you want to enable/disable IntelliCode features, you can do so through **Tools | Options | IntelliCode | General**.

IntelliCode suggestions

IntelliCode suggestions are an assisted way to carry out similar code edits in our projects. Basically, IntelliCode keeps track of the code we are writing, and if it detects code repetition that could be applied to our code, it will let us know through suggestions.

A surprising thing about IntelliCode is that it is based on the semantic structure of code, so it can also help us detect changes that we might have missed, such as changes in formulas.

For example, suppose we have some methods that allow us to calculate some static values, such as the following example:

```
public float Calculate1()
{
    var minValue = 25;
    return (float)((minValue + 126) * (Math.PI / minValue));
}
public float Calculate2()
{
    var minValue = 88;
    return (float)((minValue + 126) * (Math.PI / minValue));
}
public float Calculate3()
{
    var minValue = 56;
    return (float)((minValue + 126) * (Math.PI / minValue));
}
```

We see that the calculation follows the same structure, and the only thing that changes is the value of the `minValue` variable, so we decide to create a new method called `Calculate`, which will perform the same operation by receiving a parameter:

```csharp
public float Calculate(int value)
{
    return (float)((value + 126) * (Math.PI / value));
}
```

Subsequently, we decided to replace the code of the `Calculate1` and `Calculate2` methods to invoke the newly created `Calculate` method:

```csharp
public float Calculate1()
{
    var minValue = 25;
    return Calculate(minValue);
}
public float Calculate2()
{
    var minValue = 88;
    return Calculate(minValue);
}
```

If we go to the `Calculate3` method and click on the `return` line, an IntelliCode hint will appear, as shown in *Figure 5.16*:

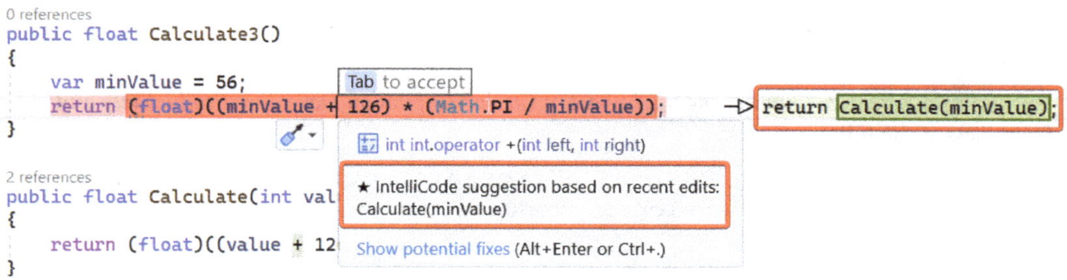

Figure 5.16 – IntelliCode suggesting an implementation of the code repetition

The suggestion made by IntelliCode tells us that we can apply the same invocation to the new method, which we can apply by pressing the *Tab* key, or we can ignore the suggestion and continue writing code.

Important note

It is very important to note that IntelliCode suggestions are only available during the development session. This means that if you restart VS, the previous hints will not appear again.

As we saw, IntelliCode provides a way to write code much faster through suggestions, which can save you several minutes a day.

Now, let's see how you can use the ChatGPT engine while developing your software projects through GitHub Copilot.

Exploring GitHub Copilot

GitHub Copilot (hereinafter referred to as **Copilot**) is a novel AI tool for programmers that can help them write code faster. Unlike IntelliCode, Copilot can take the context of comments and underlying code to supplement your code. Copilot has been trained on billions of lines of code pulled from GitHub repositories, so its knowledge of various programming languages and platforms is quite extensive.

According to the GitHub site, after conducting an analysis on the use of Copilot, it has been found that developers are up to 96% faster in repetitive tasks and that 88% of those who use it feel more productive.

Something to take into account is that the Copilot license costs 10 dollars a month or 100 dollars a year for individuals and 19 dollars for companies, although you can try it for free by starting a free trial from their website:

```
https://github.com/features/copilot
```

Note

You can find *Research: Quantifying the Impact of GitHub Copilot on Developer Productivity and Happiness* at the following link:

```
https://github.blog/2022-09-07-research-quantifying-github-
copilots-impact-on-developer-productivity-and-happiness/
```

Once you have purchased a Copilot license, you must install the **GitHub Copilot** and **GitHub Copilot Chat** extensions, as shown in *Figure 5.17*:

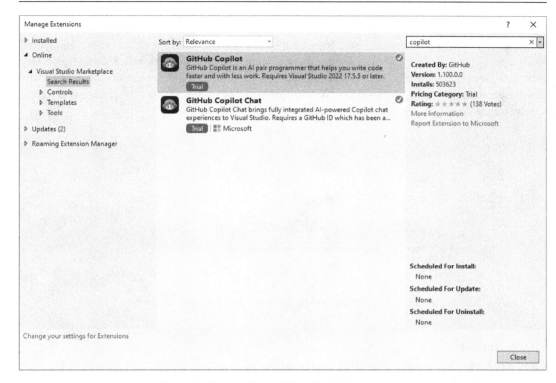

Figure 5.17 – Installing GitHub Copilot extensions

To learn more about the process of installing and searching for extensions, you can refer to *Chapter 11, Working with Extensions in Visual Studio*, where the topic is covered in depth. Now, let's move on to analyze how we can use Copilot in VS 2022.

Using Copilot for code hints

In the code repository mentioned in the *Technical requirements* section, you can find a Windows Forms project in the folder named `Chapter5_Code_Forms | Initial`, which has a form called `Form1.cs` with some controls. There's also a class called `Customer` for performing some tests, as you can see in *Figure 5.18*:

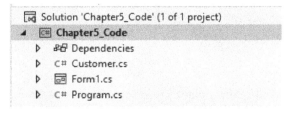

Figure 5.18 – The initial structure of the project

One way we can exploit Copilot is to get code hints. This is because it is known that some common patterns are followed while writing code. For example, if you navigate to the `Customer` class, you can see a property called `FirstName`. Suppose you want to add a property called `LastName` to the class. To achieve this with Copilot, you can add a new line after the `FirstName` property and you will immediately get the hint to create this new property, as seen in *Figure 5.19*. In case you want to complete the code, you can do it by pressing the *Tab* key.

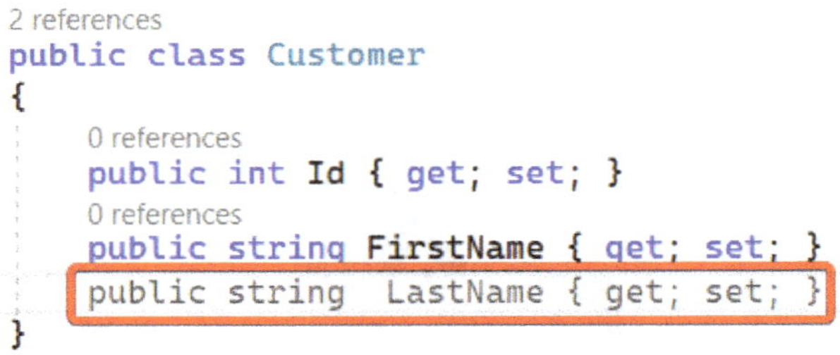

```
2 references
public class Customer
{
    0 references
    public int Id { get; set; }
    0 references
    public string FirstName { get; set; }
    public string  LastName { get; set; }
}
```

Figure 5.19 – Copilot suggesting to create the LastName property

> **Important note**
>
> It is very important to understand that the results you obtain using Copilot may vary from the results shown in the examples in this book because the output produced by the AI tools is stochastic, i.e., it may generate different variable names, methods with different implementations, and so on. Nevertheless, you should obtain results similar to those shown in the following examples.

Even if this is not the property you need, Copilot will try to help you with property suggestions—just enter the property data type and the first letter of the property to get properties that are commonly added to customer classes, as shown in *Figure 5.20*:

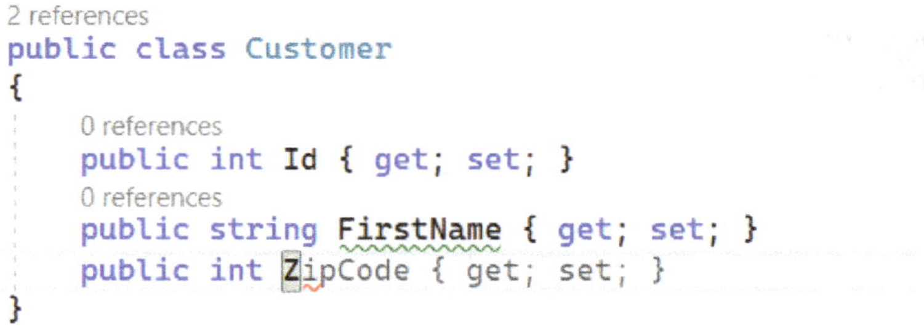

```
2 references
public class Customer
{
    0 references
    public int Id { get; set; }
    0 references
    public string FirstName { get; set; }
    public int ZipCode { get; set; }
}
```

Figure 5.20 – Copilot suggesting common properties in the Customer class

> **Important note**
>
> If Copilot has found more code hints, you can iterate through them by pressing the *Alt + .* keys.

Copilot can also help you create more complex properties. For example, if you wanted to create a property composed of `FirstName` and `LastName`, you can type the name of the property, which will cause Copilot to suggest the union of both properties in a `get`:

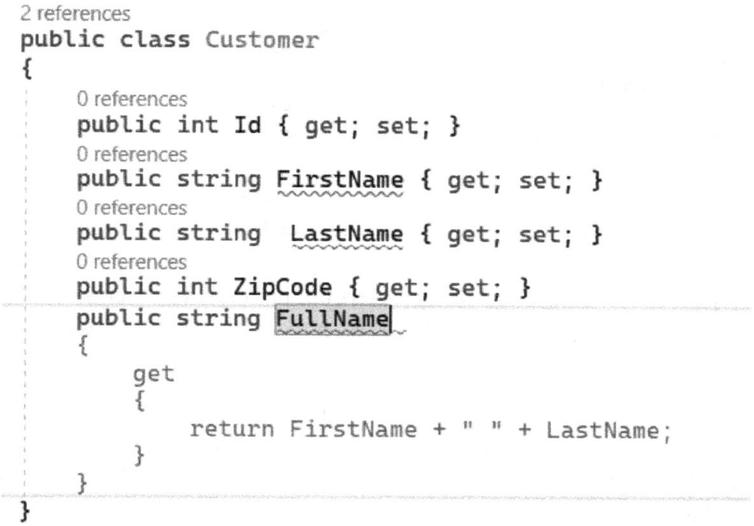

```csharp
2 references
public class Customer
{
    0 references
    public int Id { get; set; }
    0 references
    public string FirstName { get; set; }
    0 references
    public string  LastName { get; set; }
    0 references
    public int ZipCode { get; set; }
    public string FullName
    {
        get
        {
            return FirstName + " " + LastName;
        }
    }
}
```

Figure 5.21 – Copilot suggesting the code for the FullName property

Another amazing feature of Copilot is that you can instruct it to create code according to your needs through comments. For example, open the form `Form1.cs` by double-clicking it and pressing the *F7* key, which will open the code behind the form. In this class, go to the `LoadCustomers` method, and write the following comment:

```csharp
//Create a list of 5 customers whose ID starts with the number 50,
whose first names start with the letter M
```

After the comment, you will see that Copilot will show you a first suggestion to create a first customer with the data as we have indicated. After pressing the *Tab* key and creating a new line, you will see the suggestion to create the second client. Repeat until you have all five clients, as you can see in *Figure 5.22*:

```
1 reference
private void LoadCustomers()
{
    //Create a list of 5 customers whose ID starts
    //with the number 50, and whose names start with the letter M
    customers.Add(new Customer { Id = 50, FirstName = "Mary",
        LastName = "Smith", ZipCode = 12345 });
    customers.Add(new Customer { Id = 51, FirstName = "Mark",
        LastName = "Jones", ZipCode = 12345 });
    customers.Add(new Customer { Id = 52, FirstName = "Molly",
        LastName = "Smith", ZipCode = 12345 });
    customers.Add(new Customer { Id = 53, FirstName = "Mike",
        LastName = "Smith", ZipCode = 12345 });
    customers.Add(new Customer { Id = 54, FirstName = "Megan",
        LastName = "Smith", ZipCode = 12345 });
}
```

Figure 5.22 – Copilot creating new instances of the Customer class with data

You can see that Copilot has not only created customers but has been smart enough to know that we wanted to create the customers and add them to the customers field, which has saved us quite a bit of work.

Let's do a couple of additional tests. Suppose that we want to populate a CheckedListbox type control called lstCustomers in the btnLoad_Click event handler, which you can find in the Form1.cs file. To accomplish this task, we are going to place the following comment in the btnLoad_Click handler:

```
//Fill lstCustomers with the list of customers
```

Upon pressing the *Enter* key, Copilot will show you the necessary code to assign the data origin of the control to the list. In addition, Copilot will fill in the code to indicate which will be the property to show to the users. Finally, the code will be filled in to select the first customer of the list. In *Figure 5.23*, you can see the result of the execution of the program, as well as the code that Copilot has completed:

```
    customers.Add(new Customer { Id = 53, FirstName
    customers.Add(new Customer { Id = 54, FirstName
}

1 reference
private void btnLoad_Click(object sender, EventArgs
{
    //Fill lstCustomers with the list of customers
    lstCustomers.DataSource = customers;
    lstCustomers.DisplayMember = "FullName";
    lstCustomers.SelectedIndex = 0;
}

1 reference
private void btnRemove_Click(object sender, EventAr
{

}
```

Figure 5.23 – Loading customers in the form and code generated by Copilot

With this example, you have witnessed a little of the power of Copilot and how it could help you in your developments. Now, let's analyze GitHub Copilot Chat.

Using GitHub Copilot Chat

If GitHub Copilot has surprised you, imagine having it at your disposal to ask it about anything you want—code explanations, general programming doubts, and even suggestions for resolving exceptions. For this demo, we will continue working on the code behind inside Form1.cs. Suppose you want to be able to remove selected items from the list we have filled in the GitHub Copilot section. Let's use the following comment to accomplish this goal within the btnRemove_Click event handler:

```
//Remove the selected customers from CheckedListBox lstCustomers
```

With this, we get the following code that we can use in the project:

```
private void btnRemove_Click(object sender, EventArgs e)
{
    //Remove the selected customers
    //from CheckedListBox lstCustomers
    for (int i = lstCustomers.CheckedItems.Count - 1;
        i >= 0; i--)
    {
        lstCustomers
        .Items
        .Remove(lstCustomers.CheckedItems[i]);
    }
}
```

Once we have added the code in the event handler `btnRemove_Click`, we can run the application, load the customers, select one customer for removal, and press the **Remove Selected Customers** button. This causes us to obtain an exception, as shown in *Figure 5.24*:

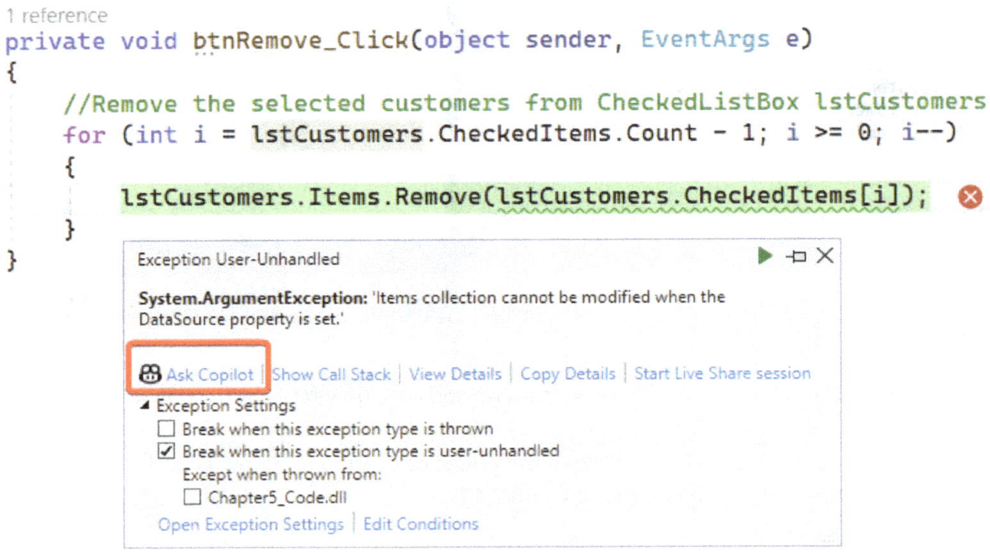

```csharp
1 reference
private void btnRemove_Click(object sender, EventArgs e)
{
    //Remove the selected customers from CheckedListBox lstCustomers
    for (int i = lstCustomers.CheckedItems.Count - 1; i >= 0; i--)
    {
        lstCustomers.Items.Remove(lstCustomers.CheckedItems[i]);  ⊗
    }
}
```

Exception User-Handled   ▶ ⊐ ✕

System.ArgumentException: 'Items collection cannot be modified when the DataSource property is set.'

🔗 Ask Copilot │ Show Call Stack │ View Details │ Copy Details │ Start Live Share session
▲ Exception Settings
  ☐ Break when this exception type is thrown
  ☑ Break when this exception type is user-unhandled
     Except when thrown from:
     ☐ Chapter5_Code.dll
Open Exception Settings │ Edit Conditions

Figure 5.24 – Exception when deleting customers and the Ask Copilot button

Perhaps, if you are an experienced programmer on the platform, you will not have any issues in solving the problem; however, how could you solve it if you have little experience or if it is the first time you encounter this error?

Fortunately, with Copilot Chat, we have available a new option that I have highlighted in *Figure 5.24*, called **Ask Copilot**, which allows us to get additional information and suggestions about the error. After pressing it, a new window will open called **GitHub Copilot chat**, where we will see information regarding the problem:

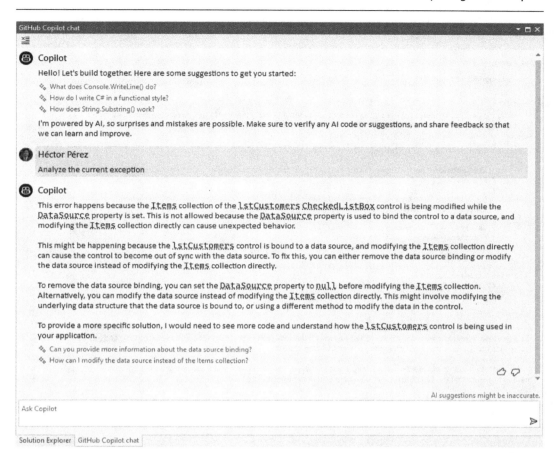

Figure 5.25 – The GitHub Copilot chat window

Inside this window, we can perform different actions from code explanations to suggested solutions to problems. For example, we see a couple of links at the end of the Copilot help that we could press and that would ask Copilot the link question. Also, we can be more specific and give Copilot instructions to help us solve the problem. In my case, I will select the **How can I modify the data source instead of the Items collection?** option to show me the code that I should place if I want to modify multiple items in the listing.

Figure 5.26 – Copilot Chat suggesting how to solve the exception.

After a few seconds, Copilot shows me the resulting code, as shown in *Figure 5.26*, which I can copy or even insert directly where the exception occurs. We can also help improve the system by evaluating how good Copilot's response was through the little thumbs up and thumbs down icons at the end of the answer provided by Copilot (see *Figure 5.26*). If we select the **Insert** button, VS will show the modification that will be carried out:

```
        private void btnRemove_Click(object sender    31      private void btnRemove_Click(object sender, Event
        {                                             32      {
            if (lstCustomers.CheckedItems.Count <=    33          if (lstCustomers.CheckedItems.Count <= 0) //
                return;                                34              return;
                                                      35
            // Remove the selected customers from     36 +        // Loop through the checked items and remove
            int lastCheckedItemIndex = lstCustomer    37 +        for (int i = lstCustomers.CheckedItems.Count
            for (int i = lastCheckedItemIndex; i >   
            {                                          38          {
                lstCustomers.SetItemChecked(lstCus    39 +            var item = lstCustomers.CheckedItems[i];
                //Modify the data source if exists    40 +            customers.Remove(item as Customer);
            }                                          41          }
                                                      42 +
                                                      43 +        // Refresh the data source with the updated
                                                      44 +        lstCustomers.DataSource = null;
                                                      45 +        lstCustomers.DataSource = customers;
        }                                              46      }
```

AI suggestions might be inaccurate.

Ask Copilot ➤   ✓ Accept   Cancel

Figure 5.27 – Preview of the changes to be made to the code

Once the code has been replaced, we can run the application again and try to remove customers. This time, the customers will be removed from the listing, as you can see in *Figure 5.28*:

Figure 5.28 – Removing customers from the list

However, when displaying the refreshed list, you can observe that the name of the class is shown and not the customer's name, which we can solve by modifying the `DisplayMember` property:

```
// Refresh the data source with the updated customers list
lstCustomers.DataSource = null;
lstCustomers.DataSource = customers;
lstCustomers.DisplayMember = "FullName";
```

This shows that although Copilot Chat is a powerful tool, it is not always perfect, so it is always advisable to verify that the outputs are correct.

Summary

VS contains a set of tools and windows that can help us a lot while we are developing our projects, as well as extensions that we can install to exploit the power of Copilot in our hands.

We have seen how CodeLens can help us find references and relationships, both through code and visually. Likewise, we studied the **Class View** window, which helps us examine class members in projects. We have also seen how IntelliCode is a new addition to the IDE, which, through AI, helps us write code quickly through various suggestions.

You have seen how GitHub Copilot can be a great ally when writing code, as it can save you many hours of development by recommending code that you can complete, as well as instructing you through comments to give you functional pieces of code. Finally, GitHub Copilot Chat is a wonderful feature that can answer almost anything—questions about general programming, the current code, and exceptions.

In *Chapter 6, Using Tools for Frontend and Backend Development*, we will look at several tools focused on web development for the development of web applications, as well as a new functionality included in VS 2022 that helps us reload a web project after making a change.

6

Using Tools for Frontend and Backend Development

Visual Studio has many tools to work with .NET applications and the Microsoft ecosystem, but it also has many tools for other programming languages and technologies. This includes web development technologies, such as JavaScript, CSS, and HTML.

In VS 2022, there are also some new improvements that help us to code faster in both the frontend and backend. This means tools for frontend developers to work with HTML, JavaScript and CSS and tools for backend developers to test web applications externally and API endpoints. With these tools, you don't need to use other editors or **integrated development environments (IDEs)** to complete your activities while working with these technologies.

In this chapter, you will learn about web tools in VS, how to take advantage of them, and how to simplify them when we are developing some common statements. These tools will help you to generate code automatically, install and specify versions of web libraries, inspect the code in JavaScript, refresh the application automatically to see the changes in real time, create dev tunnels to allow viewing the project running from other devices, and test web API endpoints using the HTTP editor.

We will learn about the following topics related to web tools:

- Using scaffolding
- Installing JavaScript and CSS libraries
- Debugging in JavaScript
- Hot Reload
- Dev tunnels
- Web API Endpoints Explorer
- HTTP editor

We will start with how to use scaffolding, which is the main tool for creating project files, with templates that are included by default in VS 2022.

Technical requirements

To complete the demos of this book chapter, you must have previously installed VS 2022 with the web development workload, as shown in *Chapter 1, Getting Started with Visual Studio 2022*.

You can obtain the chapter projects at the following link: `https://github.com/` `PacktPublishing/Hands-On-Visual-Studio-2022-Second-Edition/tree/` `main/Chapter%206`.

Using scaffolding

Scaffolding is one of the most beneficial features for developers in VS. By using scaffolding, we can save time generating code automatically by just clicking on some options.

It's indispensable to clarify that scaffolding is a popular concept in software development, and this is not unique to VS. Normally, scaffolding is associated with the code generation of **model view controller** (**MVC**) components. MVC is a popular pattern for creating web applications. Using MVC, you must distribute the responsibilities for creating web applications into three different components:

- **Model**: Responsible for saving the data
- **View**: The interface that interacts with the user
- **Controller**: In charge of handling all the actions performed by the user in the view

Let's make use of this feature by opening the initial project called `Chapter6_Code_Web`, which you can find in the repository of the *Technical requirements* section. To use scaffolding in VS, you can select the folder in the solution explorer and right-click it to open the options to select **New Scaffolded Item...** (see *Figure 6.1*):

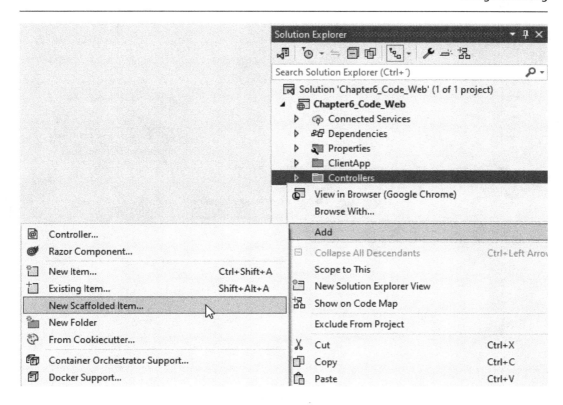

Figure 6.1 – The "New Scaffolded Item…" option in the project menu

Using this option, we have the possibility to create new elements in the project related to the MVC structure.

After clicking on **New Scaffolded Item…**, VS will provide a list of elements that we can create using the scaffolding tool:

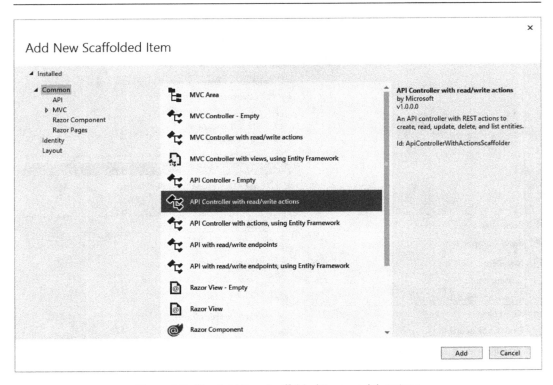

Figure 6.2 – The Add New Scaffolded Item model options

In this case, we will select **API Controller with read/write actions**, which is going to generate an API controller with the actions for the GET, POST, PUT and DELETE verbs. Choose the name GeneratedController.cs and click on **Add**:

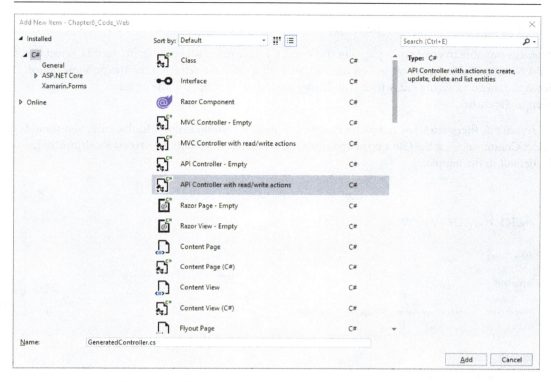

Figure 6.3 – Creating an API Controller with read/write actions

VS will generate a new controller in the `Controllers` folder with endpoints by default:

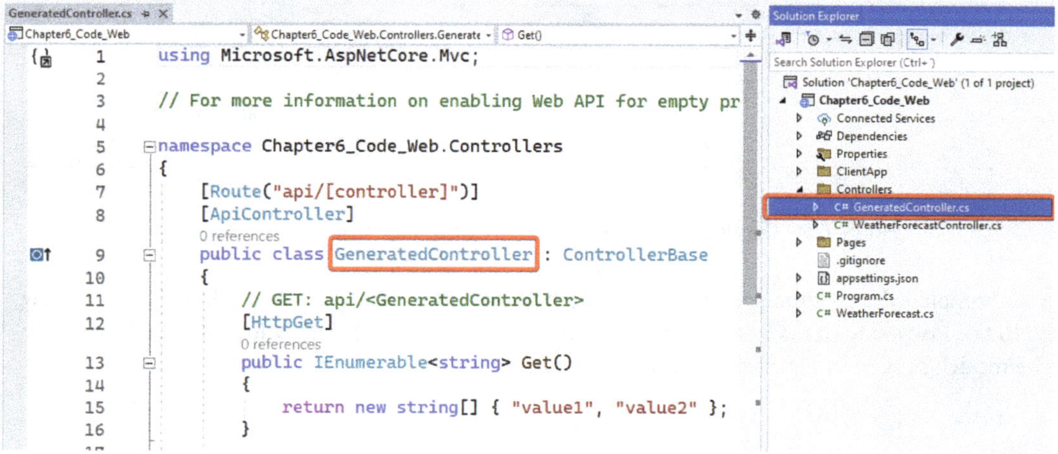

Figure 6.4 – GeneratedController is created in the Controllers folder

After creating the base template, you can replace the data type and method names to match your model.

It is also possible to create new pages in the Pages folder using scaffolding. To do this, select the **Add New Scaffolded Item** option on the Pages folder, this time selecting the **Razor View** option from the menu in *Figure 6.2*, which will display a window to configure the creation of the page, as seen in *Figure 6.5*.

In *Figure 6.5*, there are some templates that we can pick, including **Empty**. In this case, you should select **Create**, uncheck the **Use a layout page** option, and select the **WeatherForecast** model (included by default in the template):

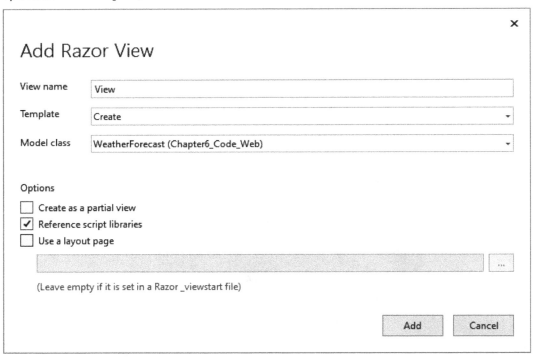

Figure 6.5 – Configuration to create a new view page using scaffolding

We can complete the creation process by clicking the **Add** button, and finally, we will see this new view page in the Pages folder. VS will analyze the model and then create a new form for each property in the model, as seen in *Figure 6.6*, considering the property:

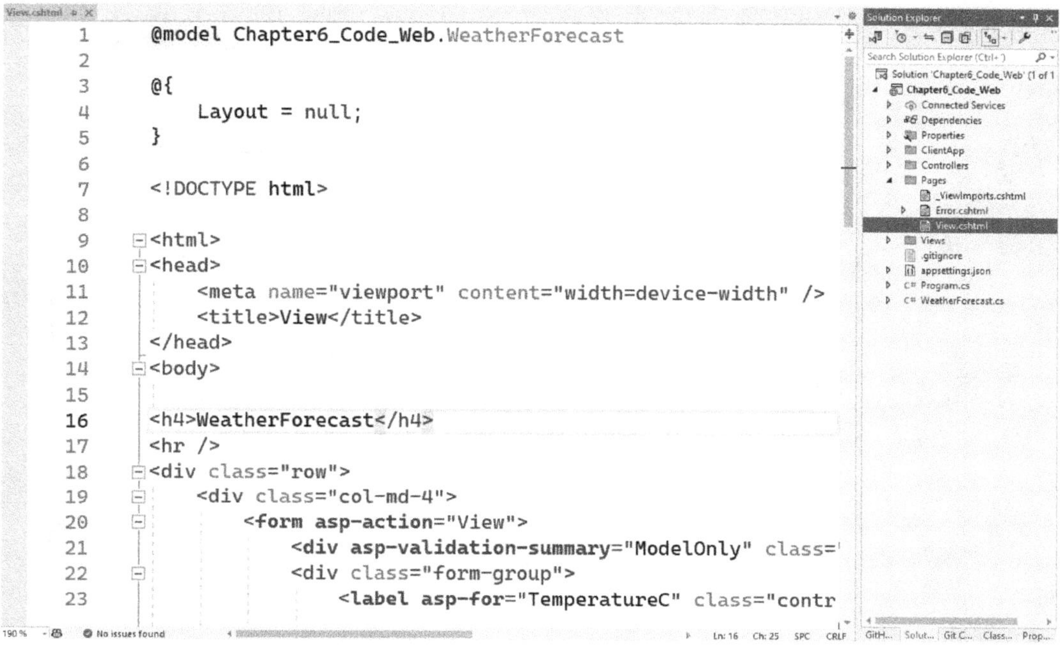

Figure 6.6 – The View.cshtml file generated from the WeatherForecast model

---

Important note

Scaffolding is related to ASP.NET code; we can create controllers and views but not JavaScript components.

Now, you know how to use scaffolding in your projects and save time using some base templates provided by VS. Let's see another tool in VS that helps us to include JavaScript and CSS libraries in our projects.

Installing JavaScript and CSS libraries

To start a project, we can use a template from VS to easily create a **proof of concept** (**POC**), demo, or base project, but there is the possibility of the project growing in terms of functionalities and services. In this scenario, we will have to include libraries to potentialize and optimize our project and extend the functions incorporated in the base template.

To include a new JavaScript library in the Chapter6_Code_Web project, you can open **Solution Explorer** and right-click on the ClientApp folder. In the menu, you will find the **Client-Side Library...** option (see *Figure 6.7*):

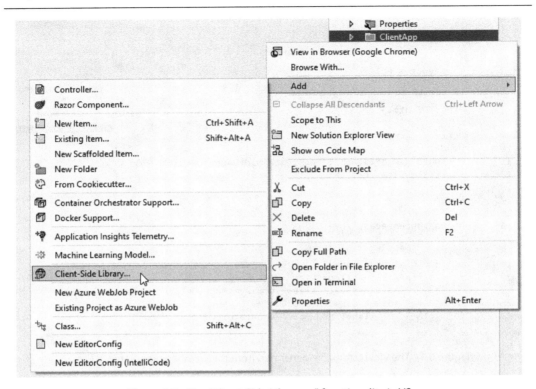

Figure 6.7 – The "Client-Side Library…" functionality in VS

After clicking on this option, you will get a model that allows you to include web libraries from different resources. By default, **cdnjs** is selected, but you can also choose the other sources supported by VS:

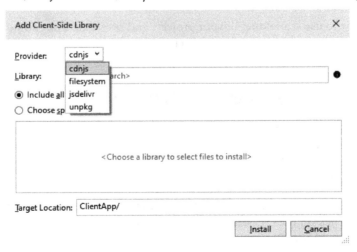

Figure 6.8 – Providers supported by VS 2022

There are three public and trusted libraries for CSS and JavaScript that we can include in our projects. VS has support for different library sources; let's review them:

- **cdnjs**: Fast and reliable content delivery for an open source library supported by Cloudflare

- **filesystem**: Custom packages in our local system

- **jsdelivr**: Free content delivery network integrated with GitHub and npm

- **unpkg**: Global open source content delivery maintained by Michael Jackson

> **Important note**
> You can create the `libman.json` file manually, include the libraries in the JSON file, and then install them using VS. For more information, go to `https://docs.microsoft.com/aspnet/core/client-side/libman/libman-vs`.

To analyze how VS adds a library to your project, you should select **cdnjs** and search by **Bootstrap**. Bootstrap is a powerful library for creating web interfaces easily using CSS classes. For more information, you can check the official documentation and quick-start guides at `https://getbootstrap.com/`.

Once you start typing, VS will suggest a list of libraries to which the written keyword is related (see *Figure 6.9*):

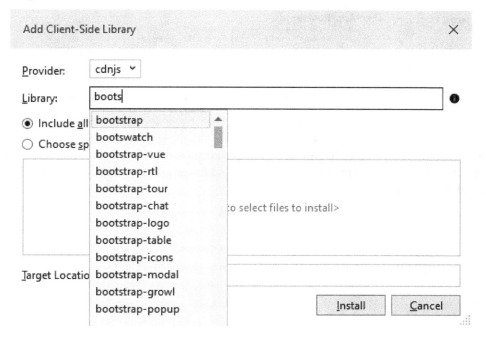

Figure 6.9 – Libraries suggested by VS

After selecting **bootstrap** (the first option recommended in the list), the most recent version of this library will be selected by default—bootstrap@5.3.2 (see *Figure 6.10*). You can choose all the components associated with the library, but normally you only use the minify version. You can select only the files that you need using the **Choose specific files** option. In *Figure 6.10*, you can see the bootstrap.min.js file, which is the only file required to use this library:

Figure 6.10 – The Bootstrap library selected and the bootstrap.min.js file picked

Now, you can click on **Install** to include Bootstrap in our project.

After installing this library, you will see a new folder that contains all the files related to Bootstrap in the ClientApp folder. Also, you will see a new file called libman.json that contains the libraries installed in the project using VS:

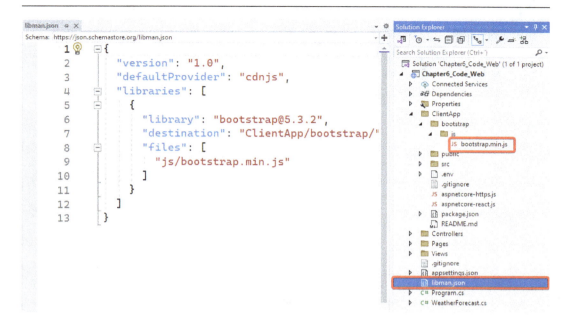

Figure 6.11 – The Bootstrap library added in the project

This file helps VS to get the libraries from the servers when the project doesn't have the files for these libraries in the repository.

VS will execute all the processes automatically and create the `libman.json` file, where we can see the version of each library and the destination folder in the project.

You now know how to include JavaScript and CSS libraries in your project using the different options supported by VS. Let's learn how to debug JavaScript code in VS to find and resolve issues quickly in the development process.

Debugging in JavaScript

We must debug a project when there is strange behavior, an issue, or a blocker in our application. VS supports debugging for many programming languages, including JavaScript. This is a great feature, given that we can debug the frontend side (for example, with JavaScript) and the backend side (for example, with C#) using the same IDE.

To debug JavaScript and TypeScript code using VS, we need to check the **Script Debugging (Enabled)** option. This option is on the execution menu of the project:

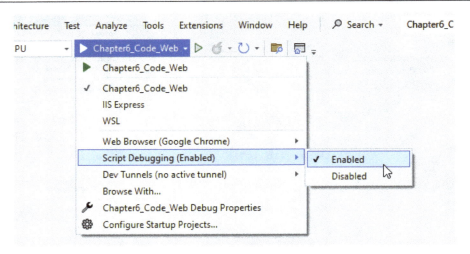

Figure 6.12 – The Script Debugging (Enabled) option in the project execution menu

Then, we can run the project in debug mode, but before that, we need to add a break to inspect the code. Navigate to `ClientApp | src | components | Counter.js` and create a new breakpoint at *line 13* (see *Figure 6.13*):

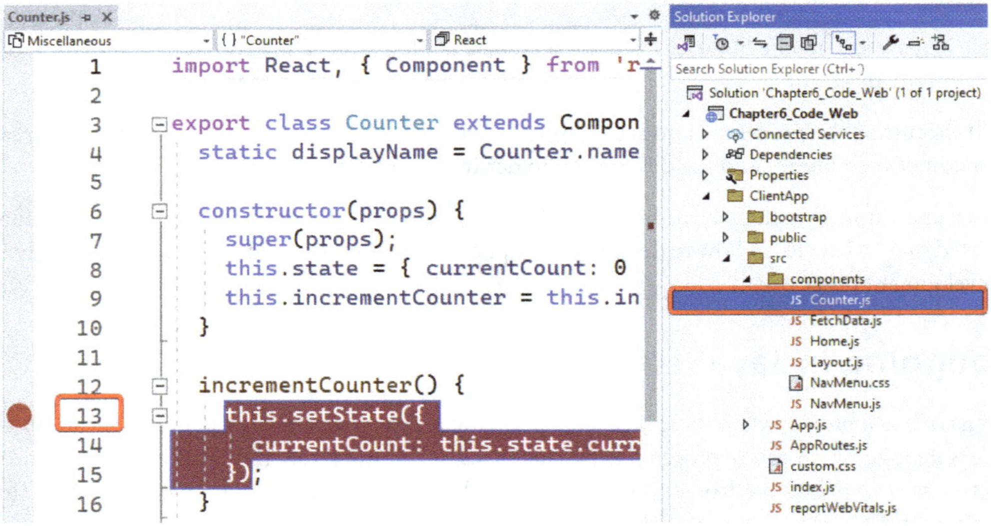

Figure 6.13 – A breakpoint in the incrementCounter method inside Counter.js

Now, execute the project using the option in the banner or press *F5* and then navigate to the **Counter** module. Once you click on **Increment**, VS will stop the execution in the `incrementCounter` method in JavaScript. In *Figure 6.14*, you can see this expected behavior:

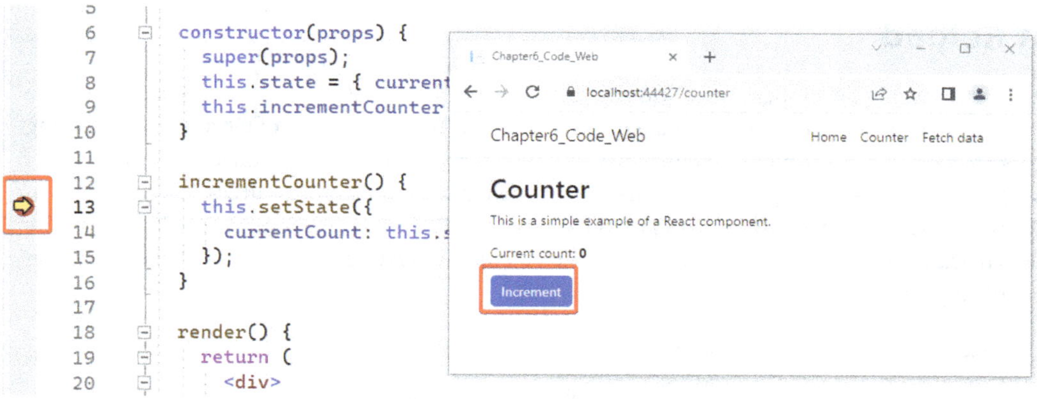

Figure 6.14 – The debugging process in the incrementCounter method with the breakpoint

At this point, we can inspect the variables in this file—for example, in *Figure 6.15*, we can see the values of `Counter.name` and `currentCount`:

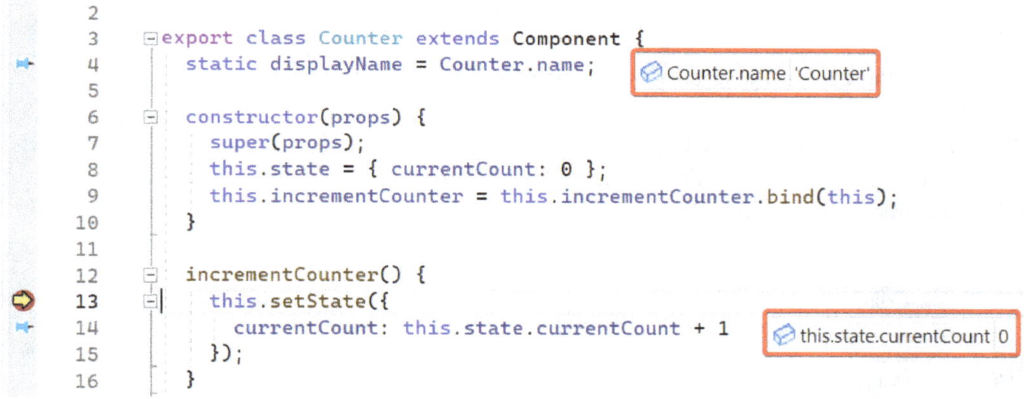

Figure 6.15 – The inspection of Counter.name and currentCount during debugging

`Counter.name` equals `Counter` and `currentCount` equals `0`. After executing the `incrementCounter` async method, the variable in the `currentCount` state will have a value of `1`.

You can now debug JavaScript code in VS and use the same tools and actions that we reviewed in *Chapter 3, Debugging and Profiling Your Apps*.

In the next section, we will review a new functionality in VS 2022 to refresh the UI after performing changes in the code.

Hot Reload

For many years, C# developers were waiting for a feature that would allow them to see real-time changes in web applications. The big challenge with this was the naturalness of C# as a programming language because C# is a compiled language. A compiled language needs to be converted to a low-level language for use by an interpreter, and this process consumes time and resources in a machine. In *Figure 6.16*, you can see a new flame-shaped icon. After clicking on this icon, you will refresh the changes in the browser, or you can select the **Hot Reload on File Save** option to reload a web application automatically after saving changes:

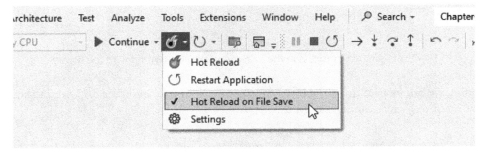

Figure 6.16 – The Hot Reload option in VS enabled during the execution

The **Hot Reload** feature has some settings that we can modify according to our needs. You can access these features using the **Settings** option when the **Hot Reload** button is enabled, or you can navigate to **Tools | Options | Debugging | .NET / C++ Hot Reload**:

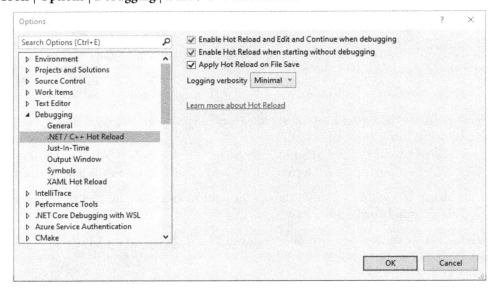

Figure 6.17 – Settings for Hot Reload in VS 2022

Let's review the options:

- **Enable Hot Reload and Edit and Continue when debugging**: This option enables Hot Reload in debug mode

- **Enable Hot Reload when starting without debugging**: This option enables Hot Reload without debugging

- **Apply Hot Reload on File Save**: After performing a change in any file and saving it, the application will reload

> **Note**
>
> If you do not enable the **Apply Hot Reload on File Save** option, you need to use the **Hot Reload** button to refresh the web application and see the changes.

To test this functionality in VS, you can run a project by pressing *F5* or using the option in the **Debug | Start Debugging** menu. After this, you can make any change in the UI—for example, you can navigate to the `NavMenu.js` file and change the name of `NavLink` from `Counter` to `Counter Module`:

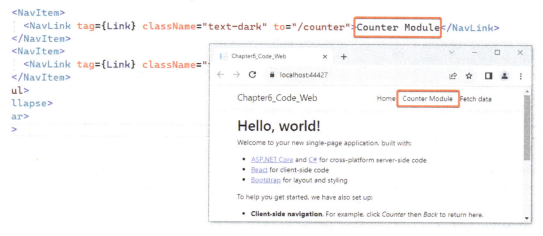

Figure 6.18 – The Counter Module NavLink in NavMenu.js

After saving the file using *Ctrl + S* or the **File | Save All** menu, you will see the change in real time in the web browser, as shown in *Figure 6.18*.

> **Important note**
>
> The wait time to see the changes in the browser depends on the project size and machine resources. Normally, it doesn't take more than two or three seconds.

We don't need to include a library in the project or install an extension in VS to use this amazing functionally. Hot reload is supported for most graphic workloads, including ASP.NET Core, WPF, .NET MAUI, Windows Forms, and others.

You can now use Hot Reload in your web project to improve your productivity when you are coding and need to test the changes in the UI quickly. Now, let's explore a new feature of VS 2022 that has arrived to change the rules of the game by allowing you to bring local environments to the cloud for testing: dev tunnels.

Dev tunnels

Dev tunnels are a feature of VS that allow you to create an ad-hoc connection to your local ASP.NET Core projects so that other devices can connect to it. For example, suppose that you create a REST service that you want to test on a mobile device, which is probably connected to a different network than localhost. Then, through a dev tunnel, you can create a communication tunnel between the REST API and the device.

Another use case is if you would like to make an online presentation of a project, and you would like the attendees to be able to observe the application being executed. These are just a few use cases that you could use dev tunnels for. Let's see how to create and use them.

Creating a dev tunnel

To follow this demonstration, you can create a new project with the `ASP.NET Core Web App` template or use the project named `Chapter6_Code`, which you can download from the link indicated in the *Technical requirements* section.

Once the project is open, go to the options button with the **https** text and select the **Dev Tunnels (no active tunnel)** | **Create a tunnel** option, as you can see in *Figure 6.19*:

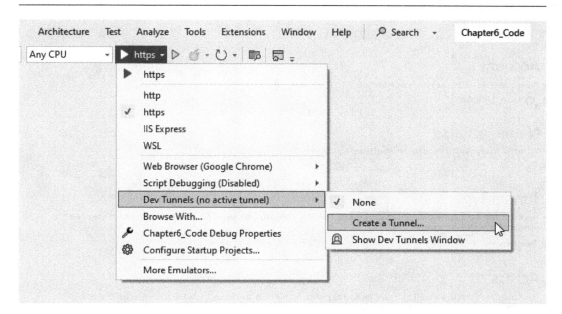

Figure 6.19 – Option to create a dev tunnel

Once you have clicked on the **Create a tunnel** option, a new window will open, where you will need to fill in the following information:

- **Account**: This is an Azure, Microsoft, or GitHub account that will be used to create the tunnel.

- **Name**: This is the name type identifier that will be reflected in the VS interface.

- **Tunnel Type**: This is the type of tunnel to create, either **Persistent** or **Temporary**. The term "persistent" does not mean that the tunnel will remain open as long as VS is closed; rather, it means that the URL assigned to the tunnel will be the same each time VS is started. On the other hand, a temporary tunnel gets a new URL each time VS is restarted.

- **Access**: The type of authentication for the tunnel, which can be of three types:

  - **Private**: It is only accessible by whoever created it.

  - **Organization**: Accessible to accounts within the same organization as the tunnel creator.

  - **Public**: Anyone with the URL can access the web app or API.

For our example, we are going to create a tunnel with the name VS_Tunnel of the **Temporary** type with **Public** access, as shown in *Figure 6.20*:

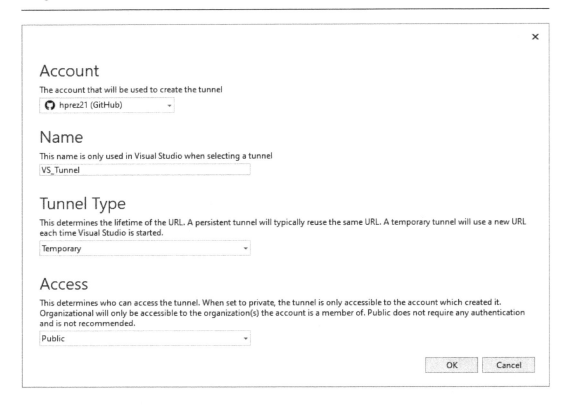

Figure 6.20 – Configuring the dev tunnel

When the tunnel is created, you will see a message indicating that the tunnel has been successfully created and that it has been selected as the current active tunnel:

Figure 6.21 – Message indicating successful creation of the dev tunnel

In case you want to validate whether there are any open tunnels, see which tunnel is currently selected, or even select a different tunnel, you can do so by checking the **https | Dev Tunnels (VS_Tunnel)** options again. In *Figure 6.22*, you can see that the **VS_Tunnel** has been selected after it has been created:

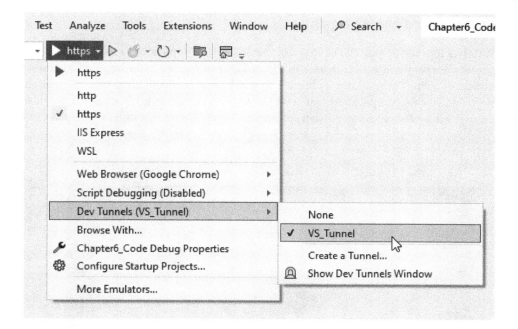

Figure 6.22 – Option showing the selected dev tunnel

VS also has a window dedicated to managing the dev tunnels you have created, which you can access through **View | Other Windows | Dev Tunnels**. In this window, you will be able to find the dev tunnels previously created, the option to create new dev tunnels, show the output logs, and see the dev tunnels usage limits.

On the other hand, if you right-click on each Dev Tunnel, you will be able to perform actions such as enabling or disabling the dev tunnel, copying the dev tunnel token and deleting the dev tunnel, as you can see here:

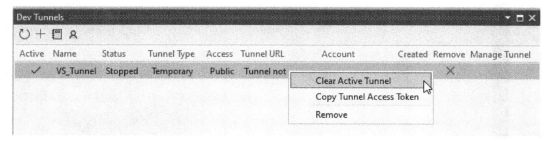

Figure 6.23 – Dev Tunnels window

Once we have created the tunnel, let's see how to connect through it.

Connecting through the tunnel

The next step is to test the tunnel. For this, you must start the application execution, which will open a new window of the selected browser. In *Figure 6.24*, you can see the result of the execution in which, instead of seeing the resulting application in localhost as normally happens, a message about the connection to the dev tunnel appears in a URL with the ending `devtunnels.ms`. By clicking **Continue**, you will be redirected to the web application:

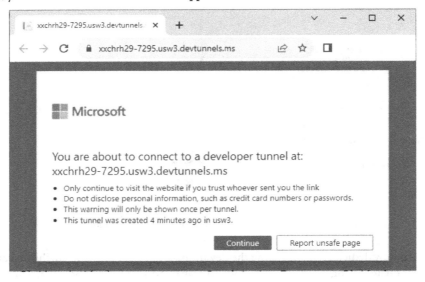

Figure 6.24 – Running the project with a URL using the dev tunnel

You may not think this is a big change; however, if you go to a different device and enter the generated URL, you will be able to access the resulting web application, as you can see in *Figure 6.25*:

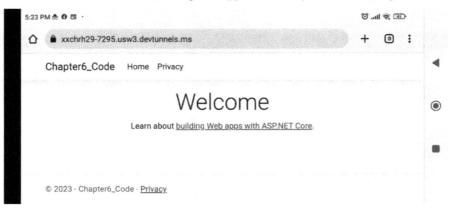

Figure 6.25 – The application from a mobile device on a different network

Without a doubt, dev tunnels are a powerful feature allowing other devices to connect to our ASP.NET Core applications or web APIs. Now, let's see how Web API Endpoints Explorer can help us create APIs faster.

Web API Endpoints Explorer

Web API Endpoints Explorer is a VS tool that allows interaction with API projects based on ASP.NET Core. For this demonstration, from the resources you downloaded in the *Technical requirements* section, you have available a project named `Chapter6_Code_API` to which I have added several endpoints.

Once you have opened the project, you can test the endpoints explorer through **View | Other Windows | Endpoints Explorer**, which will display a new window called **Endpoints Explorer**, as you can see in *Figure 6.26*:

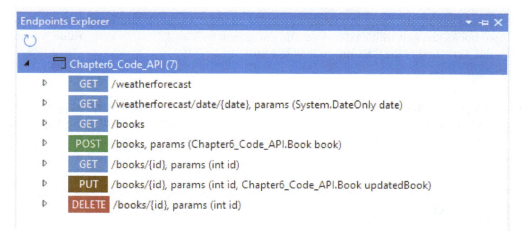

Figure 6.26 – The Endpoints Explorer window

The **Endpoints Explorer** displays all the endpoints it has found in your project, as well as the type of HTTP method and its path. If you right-click on one of the endpoints, you will be presented with two options:

- **Open in the Editor**: This action will open the controller where the endpoint is specified
- **Generate Request**: This action will generate an `http` file and add it to the project

Let's put the Endpoints explorer to the test. To do so, follow these steps:

1. Start the execution of the project.
2. Open **Endpoints Explorer**.

3. Right-click on the Endpoint **GET /books/{id}, params (int id)**, and select the **Generate Request** option. This will generate a `.http` file, such as the one shown in *Figure 6.27*:

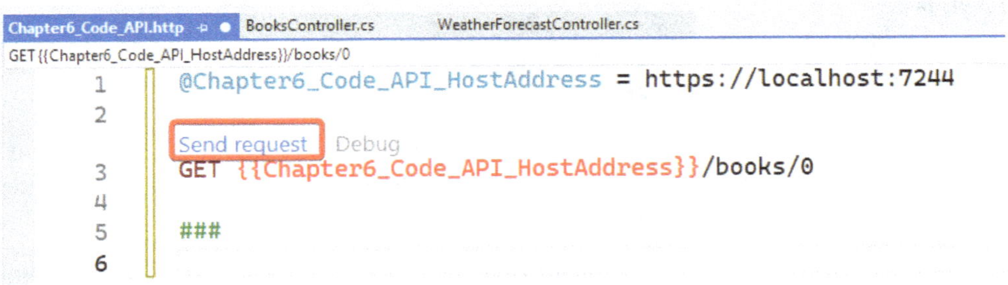

Figure 6.27 – .http file created with a Get request with the option to send the request

4. You can modify the file to fit a valid request. For example, the book controller creates a list of two books, the first with an `Id = 1` and the second with an `Id = 2`, so the value 0 is not valid. Let's modify the GET line to look like this:

```
GET {{Chapter6_Code_API_HostAddress}}/books/1
```

5. Once you have modified the file and saved the changes, click on the **Send Request** option that I have marked in *Figure 6.27*. This will open a new window showing the result of the execution. In this case, this is a status of 200 indicating a correct response:

Figure 6.28 – Get request response with a status of 200

Without a doubt, Endpoints Explorer is a tool that will help you a lot when testing the endpoints of your APIs. Now, let's take a closer look at the HTTP editor.

HTTP editor

The .http file editor is a friendly way to test mainly API projects created with ASP.NET Core. Some of the features of this editor are as follows:

- It allows the creation and update of .http files

- It allows the testing of HTTP requests specified in the .http file

- It displays responses to the requests

To better understand the syntax within an .http file, let's continue working with the Chapter6_Code_API project. This time, create a request for the **POST /books endpoint, params (Chapter6_Code_API.Book book book)**, as you can see in *Figure 6.29*:

```
1    @Chapter6_Code_API_HostAddress = https://localhost:7244
2
     Send request | Debug
3    GET {{Chapter6_Code_API_HostAddress}}/books/1
4
5    ###
6
       Send request | Debug
7    POST {{Chapter6_Code_API_HostAddress}}/books
8    Content-Type: application/json
9
10   {
11     //Book
12   }
13
14   ###
15
```

Endpoints Explorer

▲ Chapter6_Code_API (7)
  ▷ GET /books
  ▷ POST /books, params (Chapter6_Code_API.Book book)
  ▷ GET /books/{id}, params (int id)
  ▷ PUT /books/{id}, params (int id, Chapter6_Code_API.Book updatedBook)
  ▷ DELETE /books/{id}, params (int id)
  ▷ GET /weatherforecast
  ▷ GET /weatherforecast/date/{date}, params (System.DateOnly date)

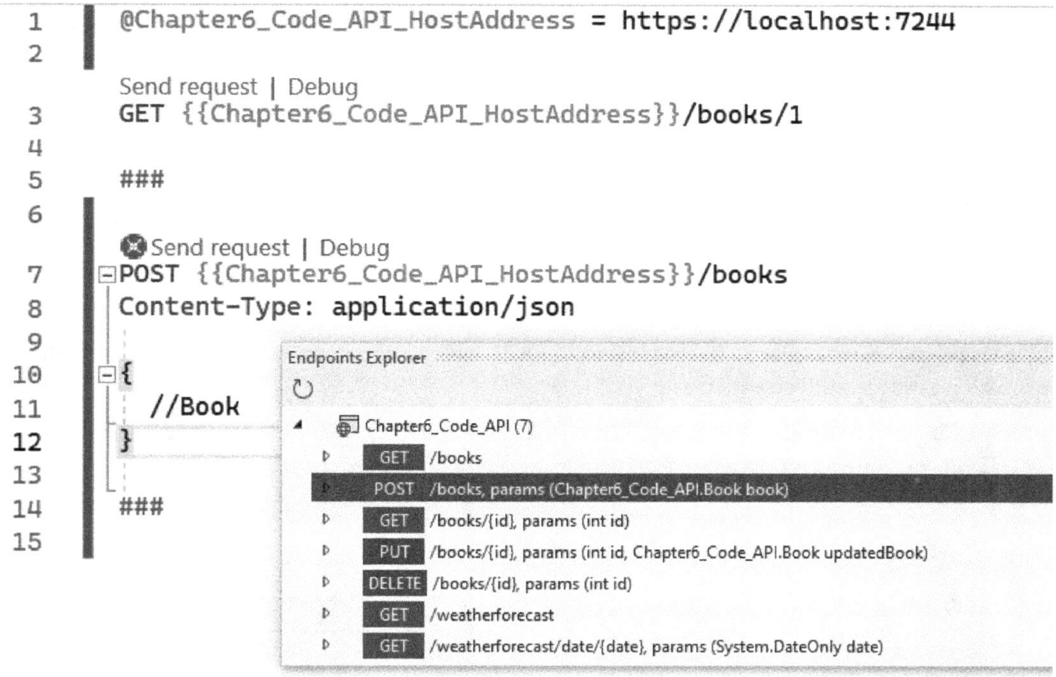

Figure 6.29 – POST request added according to the Endpoint selected

In case there is a previously created .http file, it will be updated to display the new HTTP request, as you can see in *Figure 6.29*.

Let's analyze the syntax of the file:

- *Line 1*: A variable is created, which we can identify by the '@' symbol, called `Chapter6_Code_API_HostAddress`. The variable has a value equal to the main address of the service. This is for the purpose of being able to change the value in one place in case we need it. You can change this name to a shorter one if you prefer.

- *Lines 3* and *7*: The type of HTTP request and its endpoint are defined. You can see that the variable defined in line 1 is reused, using the name between a pair of double curly braces as follows: `{{variable_name}}`.

- *Lines 5* and *14*: The `###` structure is used to delimit the end of an HTTP request.

- *Line 8*: Since this is a `POST` request, the type of content that will be sent through the HTTP request is specified.

- *Lines 10 - 12*: This is the body for the `POST` request. VS has placed a comment indicating that we need to fill in the information for a new book to be added, so we can use the following JSON string for this purpose:

```
{
    "id": 5,
    "title": "Moby Dick",
    "author": "Herman Melville",
    "genre": "Adventure"
}
```

With these changes, we can send the request to the API, which will return a `201` code, meaning that the record was successfully created, as you can see in *Figure 6.30*:

Figure 6.30 – Book object added to the POST request with a 201 response

The HTTP editor is an excellent way to make requests to your ASP.NET Core-based APIs without the need for external tools. This brings us to the end of this chapter. Let's now review what we learned.

Summary

Now, you are ready to take advantage of the web tools in VS to code faster and improve the quality of your code. With scaffolding, we can create components for an MVC model easily; VS generates the code using a template with simple sample code.

You also learned how to include JavaScript and CSS libraries using the tools included in VS. Using these tools, you know how to select the right version of the library and upgrade dependencies in the future.

If there is an issue or strange behavior in the code, you can now use JavaScript debugging to analyze the code deeper and execute the code step by step, inspecting the values of the variables and workflows.

Also, you learned how to use Hot Reload in VS 2022 to refresh the application when you are debugging and see the changes performed in the code in real time.

Another tool you have learned to use is dev tunnels, which are a way to test ASP.NET Core-based applications and APIs from other devices even if they are on different networks, which can make testing easier.

Finally, you have seen that by combining the Web API Endpoints Explorer and the HTTP Editor, it is very easy to test API Endpoints without the need to use external tools and with the ease of maintaining an .http file.

In *Chapter 7, Styling and Cleanup Tools*, we will continue learning about tools included in VS that improve our experience working with styles and CSS. We will also use some tools to clean up code by choosing a specific file or a whole project.

7

Styling and Cleanup Tools

Whether you are a frontend developer or a backend developer, having tools that help you maintain clean code is essential to move projects forward in an efficient manner.

Similarly, having tools that allow you to edit CSS files quicker helps the development team move faster on a project. Fortunately, VS has several tools that you can use while working with CSS files that will allow you to write and complete your styles quickly and in a user-friendly way.

If you are a backend programmer and you use C# or Visual Basic, you should also know that there are code analysis tools, both to maintain good quality and to follow nomenclatures that you can define.

Here are the main topics we will cover in this chapter:

- Working with CSS styling tools
- Cleaning code with code analysis tools

Technical requirements

To perform the tests that will be explored throughout this chapter, you must have installed the workload shown in *Chapter 1, Getting Started with Visual Studio 2022*.

In addition, to follow along with the *Working with images* section, you must install an additional component called **Image and 3D model editors**, as shown in the following figure:

Figure 7.1 – The "Image and 3D model editors" individual component selection

You can find the example source code in this book's GitHub repository: `https://github.com/PacktPublishing/Hands-On-Visual-Studio-2022-Second-Edition/tree/main/Chapter%207`.

Woking with CSS styling tools

Having tools for editing CSS files is an advantage for frontend web developers as it allows them to edit these files quickly and easily. That is why VS contains several tools that can be of great help in creating and editing these files.

We'll start by examining CSS3 snippets.

CSS3 snippets

Even today, there are cross-browser compatibility issues for displaying styles. Surely, it must have happened to you that when implementing a CSS property; it looks different on each browser.

It is for this reason that VS has implemented a CSS3 snippet completion system that allows cross-browser compatibility without the need to write code for each browser.

To see this practically, we can open the `Chapter7_Code | wwwroot | css | app.css` file and locate the `.content` style. Within this style, we can start typing `border-radius`. This will display a list of IntelliSense recommendations, as shown in *Figure 7.2*:

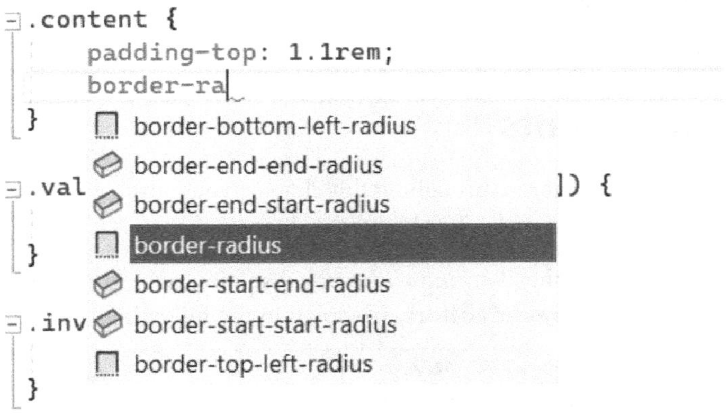

Figure 7.2 – Recommendations for the border-radius term

As you can see, there are two types of icons in this list, some with a blue geometric shape and others with a square white figure. Among these icons, we are interested in the white ones, since they are the CSS3 snippets. We can scroll through the list with the up and down keys on the keyboard. Once we have selected the snippet we are interested in, we can press the *Tab* key twice. This will result in the implementation of the cross-browser-compatible CSS3 snippet, as shown in the following code block:

```
.content {
    padding-top: 1.1rem;
    -moz-border-radius: inherit;
    -webkit-border-radius: inherit;
    border-radius: inherit;
}
```

Among the most common multi-browser CSS3 styles, we can find the following:

- **Background styles**: We can use this to set the background appearance of the elements
- **Border styles**: These allow us to set a border on the elements
- **Flex styles**: These involve the use of the Flexbox CSS layout
- **Grid styles**: These allow the CSS grid to be designed and position elements in rows and columns
- **Text styles**: These set typeface characteristics such as font family, size, weight, and so on

As you can see, CSS3 snippets can help you create styles by effectively attacking cross-browser compatibility.

Now, let's see how VS can help us understand CSS styles more quickly through indentation.

Hierarchical CSS indentation

Style indentation is a visual aid that can increase productivity considerably by showing the content of a style through spaces at the beginning of a line, as well as the sub-styles belonging to a parent style.

VS allows you to create a quick indentation in the style files. Suppose, for example, you want to create a style called .main and a sub-style that affects all div elements within the .main style, as shown in the following code block:

```
.main {
    padding: 0px 12px;
    margin: 12px 8px 8px 8px;
    min-height: 420px;
}
.main div {
    border: 25px;
}
```

In principle, if you have written the styles at the same indentation level, you can apply the indentation by navigating to the **Edit | Advanced | Format Document** menu to perform hierarchical indentation of the whole document, as shown in the following code block:

```
.main {
    padding: 0px 12px;
    margin: 12px 8px 8px 8px;
    min-height: 420px;
}
    .main div {
        border: 25px;
    }
```

If, on the other hand, you only want to apply the indentation to a specifically selected set of styles, you can select the **Edit | Advanced | Format Selection** option.

> **Note**
>
> It is possible to customize the indentation values through the **Tools | Options | Text Editor | CSS | Tabs** option.

Now, let's look at the color picker feature in VS.

Color picker

One feature that is extremely useful when working with styles is the ability to select a color for an element. Fortunately, VS has a built-in color picker that, although looking very simple, does its job very well.

To test it, let's edit the `.btn-primary` style we created in the *Hierarchical CSS indentation* section. Type the `background-color:` attribute; this will show you a vertical display of predefined colors with an assigned name that you can select. Ignore this list and type the # symbol instead. Immediately, you will see a new horizontal list of predefined colors, as shown in *Figure 7.3*:

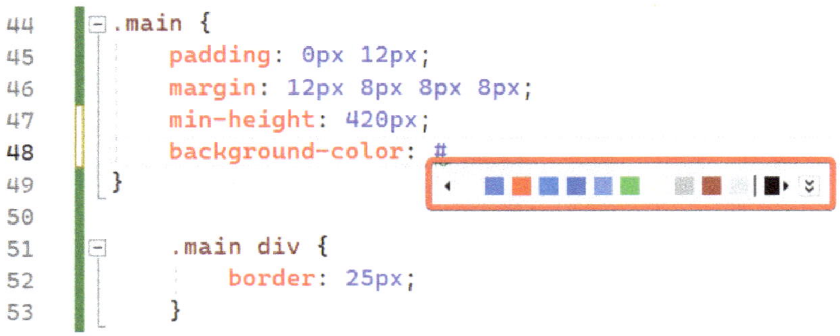

Figure 7.3 – A horizontal list of predefined colors

If you want to set a custom color, you can click on the button at the end of the color list, which will display the color picker, as shown here:

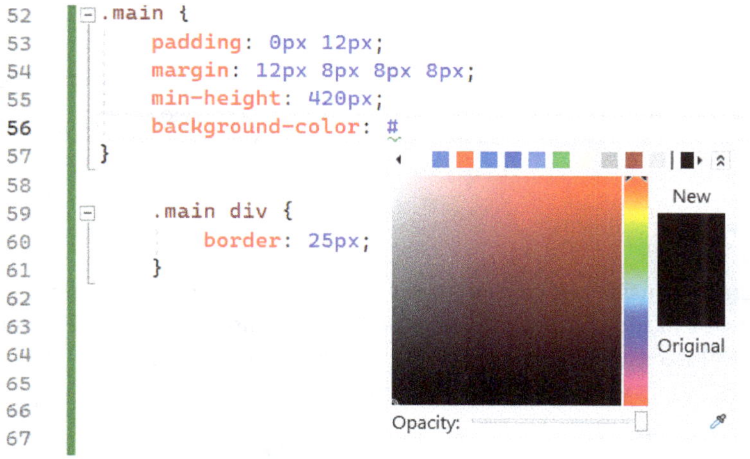

```
52   .main {
53       padding: 0px 12px;
54       margin: 12px 8px 8px 8px;
55       min-height: 420px;
56       background-color: #
57   }
58
59   .main div {
60       border: 25px;
61   }
62
63
64
65
66
67
```

Figure 7.4 – The CSS color picker

This way, it is possible to select a color from the color selection, change the color hue, add opacity or transparency to the selected color, and even use the eyedropper tool to select the color from an external source, such as an image. For this demonstration, I have selected the color with the #1b0b8599 code, as seen in *Figure 7.5*:

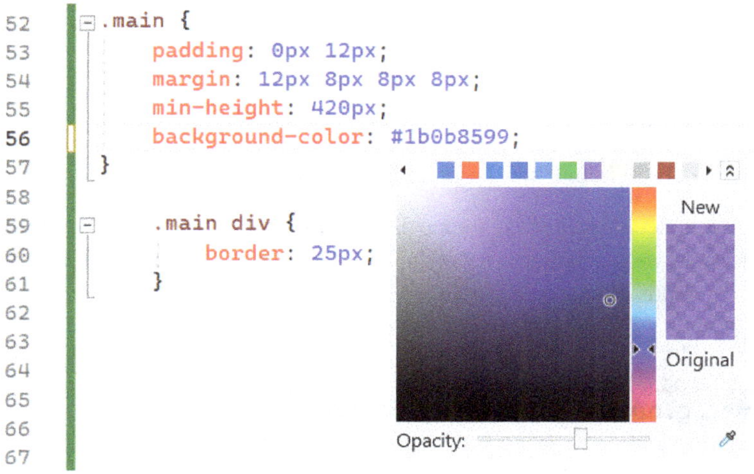

```
52   .main {
53       padding: 0px 12px;
54       margin: 12px 8px 8px 8px;
55       min-height: 420px;
56       background-color: #1b0b8599;
57   }
58
59   .main div {
60       border: 25px;
61   }
62
63
64
65
66
67
```

Figure 7.5 – Selecting a custom color from the color picker

Without a doubt, the color picker is a tool that can help us a lot when we need to assign a custom color. Now, let's learn how IntelliSense can help us write faster in CSS files.

IntelliSense in style files

Just as IntelliSense can be an extraordinary help when creating source code, it can also be very useful when creating style files.

Let's look at some practical examples of IntelliSense usage. Let's go to the Chapter7_code | wwwroot | css | app.css file. Inside this file, let's proceed to create a new style called .intellisense, as shown in the following code block:

```
.intellisense{
}
```

If we position ourselves inside the style and press the *Ctrl + spacebar* key combination, a list of all the attributes that we can add to the newly created style will be displayed. If we start typing the name of an attribute, it will start filtering the list with the matches of what we type, as shown in *Figure 7.6*:

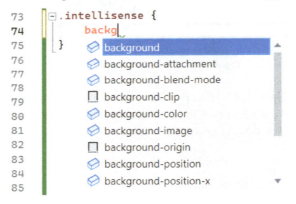

Figure 7.6 – IntelliSense showing recommendations

It is also possible to select an item from the list and complete the item name by pressing the *Tab* key.

For this demonstration, let's assume that we need to use the background attribute, but we do not know the possible values we can assign to it. IntelliSense can help us visually by showing us an example of the use of each attribute, as shown in the following figure:

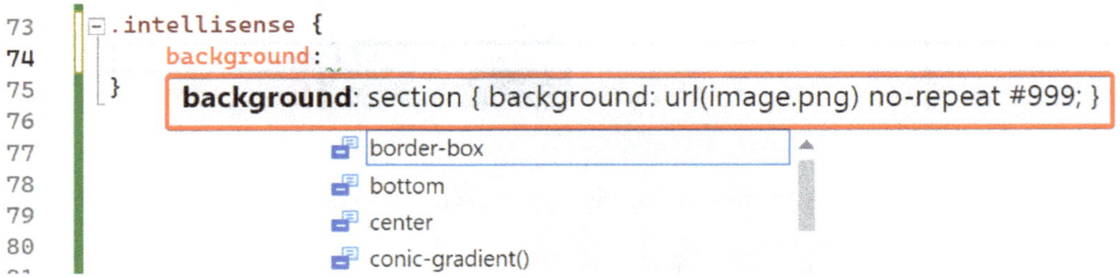

```
73   .intellisense {
74      background:
75   }
76
77      border-box
78      bottom
79      center
80      conic-gradient()
```

background: section { background: url(image.png) no-repeat #999; }

Figure 7.7 – IntelliSense showing a possible use of the background attribute

Not only that, but IntelliSense also adapts the results to the context of the selected attribute. For example, suppose we need to assign a set of fonts to the `font-family` attribute of a style. If we type the `font-family` attribute, VS will provide us with the list of values corresponding to the `font-family` attribute, as shown in *Figure 7.8*:

Figure 7.8 – Recommended values for the font-family attribute by IntelliSense

If, on the other hand, we want to assign a value to the `font-weight` attribute, we will see results according to this attribute, as shown in *Figure 7.9*:

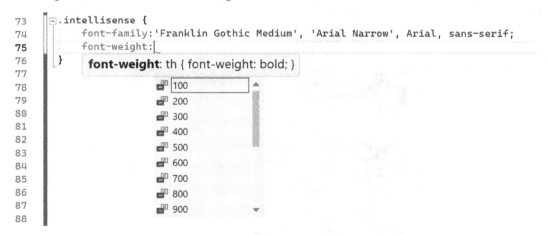

Figure 7.9 – Recommended values for the font-weight attribute by IntelliSense

Undoubtedly, IntelliSense is an excellent aid for the creation of styles. Now, let's move on to analyze the image editor.

Working with images

A tool that is not very well known in VS is the **image editor**. This tool must be installed as specified in the *Technical requirements* section, and without a doubt, it can help us with editing the images of our project.

Here are some of the situations in which the image editor is useful:

- When we need to rescale an image

- When we need to change the color of a section to another color

- When we need to rotate an image

- When we need to add text to an image

- When we need to apply a filter to an image

In the repository mentioned in the *Technical requirements* section, I added an image located at `Chapter7_Code | wwwroot | visualstudiologo.png` to perform different tests with the image editor.

Once we open an image (in this case, `visualstudiologo.png`), we will see two toolbars – one located on the left-hand side, called the *image editor* toolbar, and the second one on the top, called the *image editor mode toolbar*, as shown in the following figure:

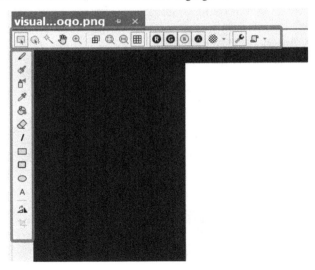

Figure 7.10 – The VS image editor

First, let's analyze the image editor toolbar. This is a bar that appears on the left-hand side of the editor and contains tools that allow you to perform some action on the image, such as adding geometric shapes or rotating the image.

At the top, we have the image editor mode toolbar. This toolbar contains buttons that execute advanced commands, such as *irregular selection*, *wand selection*, *pan*, *zoom*, and *image properties*.

Let's look at a practical example. Let's assume that we need to execute the following tasks on the image:

- Convert the image into grayscale

- Flip the image horizontally

- Write the text `Visual Studio Logo` on the image

To execute these tasks, we must perform the following steps in order:

1. In the image editor mode toolbar, select the **Advanced | Filters | Black and White** option, as shown in *Figure 7.11*:

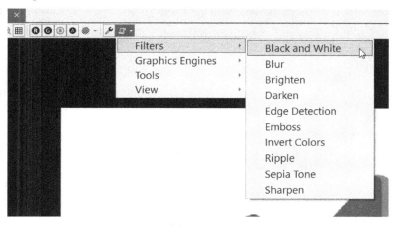

Figure 7.11 – Converting the image into grayscale

2. In the image editor toolbar, double-click the *rotate image* button, as shown in *Figure 7.12*:

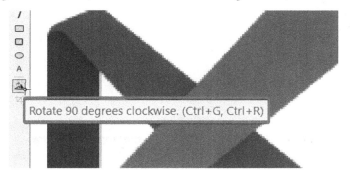

Figure 7.12 – Rotating the image

3. Select the *text* tool, as shown in *Figure 7.13*:

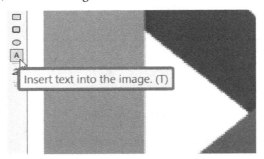

Figure 7.13 – Selecting the text tool

4. Add the text `Visual Studio Logo` in the **Properties** window, as shown in *Figure 7.14*:

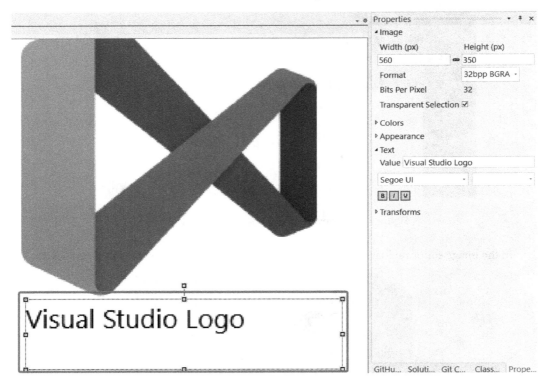

Figure 7.14 – Changing the Text value in the Properties window

With these edits applied, we will have the result shown in *Figure 7.15*:

Visual Studio Logo

Figure 7.15 – The result after applying the changes

The last thing to do is save the image so that the changes are permanently applied to it. With this example, we have seen how the image editor can be very useful if we need to make edits to our images.

In the next section, we'll understand how code analysis can help .NET developers have clean and quality code.

Cleaning code with code analysis tools

VS 2022 includes a series of C# or Visual Basic code analyzers that allow us to maintain good code quality and consistent style in the source code. To use this feature by default, projects must be configured on a framework version of .NET 5 or higher. To differentiate compilation errors, analysis violations will appear with the CA prefix in the case of a code quality analysis violation and IDE in the case of a style analysis violation.

The code analysis tools correspond to code quality analysis and code style analysis, so we will see what they are, how they can help us, and learn how to set up and run code cleanup profiles. First, let's look at how to take advantage of code quality analysis.

Code quality analysis

Code quality involves having secure source code, with the best possible performance and good design, among other characteristics. Fortunately, VS can help us to maintain high-quality code through rules enabled by default.

To visualize one of these violations practically, let's go to the Program.cs file and add the following line at the end of the file:

```
int value1 = 1;
int value2 = 1;
Console.WriteLine(Object.ReferenceEquals(value1, value2));
```

Now, to compile the project, right-click on the project name and click on the **Build** option, as shown in *Figure 7.16*:

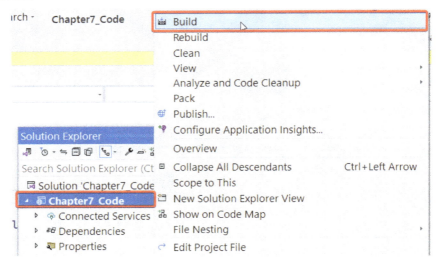

Figure 7.16 – Compiling the project

When performing the compilation, we won't see errors immediately; however, if you go to the **Error List** tab and see the list of **Warnings**, you will see some marked with the CA prefix. In our specific example, we can see the **CA2013** warning, as shown in *Figure 7.17*, which tells us not to pass a value of the int type to the ReferenceEquals method because it will always return a *false* value due to the *boxing* operation (conversion from a *value* type to a *reference* type) of the value:

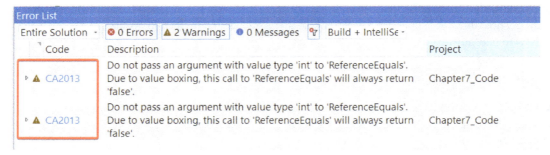

Figure 7.17 – A code quality warning

> **Important note**
>
> Occasionally, VS may suggest code fixes to fix warnings in the code, through a light bulb icon appearing on the error. You can find the complete list of code quality rules at `https://github.com/dotnet/roslyn-analyzers/blob/main/src/NetAnalyzers/Core/AnalyzerReleases.Shipped.md`.

Now, let's look at the rules that are applied to the code style.

Working with code styles

Code styles are a configuration that can be quite useful for C# and Visual Basic developers to keep a project with correct nomenclature, especially if the project is used by several members of a team.

Code styles can be created for a specific project or VS instance installed on a machine.

You can use code styles by opening the **Tools | Options | Text Editor | C#** section or **Visual Basic | Code Style | General** section:

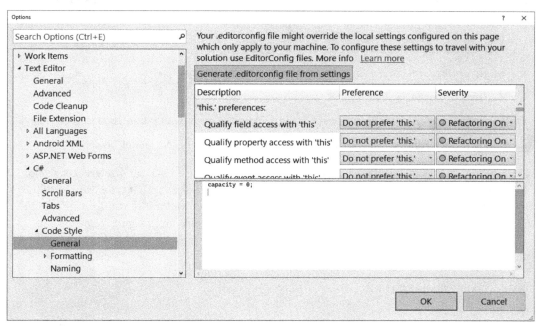

Figure 7.18 – The code style configuration window

Once we are in this window, we will be able to see the code style configuration for the current machine. We can change any of the given options to fit the code nomenclature we need. These options are divided into the following categories:

- **this preferences**: Preferences for the `this` prefix, when accessing class members
- **Predefined type preferences**: Preferences for whether to use predefined types such as `int` or `string`, or framework types such as `Int32` or `String`
- **var preferences**: Preferences for the use of the `var` keyword, especially if it should be changed to the specific type that is declared
- **Code block preferences**: Code formatting preferences, such as the use of top-level statements
- **Parentheses preferences**: Preferences for the use of parentheses in expressions for clarity of operations
- **Expression preferences**: Style preferences for expressions, such as syntax for object initializers, Boolean expressions, expression body, and others
- **Pattern matching preferences**: Preferences for pattern matching, such as the use of `is` over `' == '`
- **Variable preferences**: Preferences for variable declaration
- **'null' checking preferences**: Preferences for `null` checks
- **'using' preferences**: Preferences for the use of `using` directives
- **Modifier preferences**: Preferences to access modifiers
- **Parameter preferences**: Preferences for the declaration and use of parameters in methods
- **New line preferences**: Preferences on the insertion of new lines for better readability

To give you a quick idea of how these preferences can affect your code, in the **'var' preferences** section, we have the **For built-in types** option, which has two options: **Prefer explicit type**, which will not warn you about the data types you declare, and **Prefer 'var'**, which will notify you according to the indicated severity type when you use the `var` keyword. This can be seen in *Figure 7.19*:

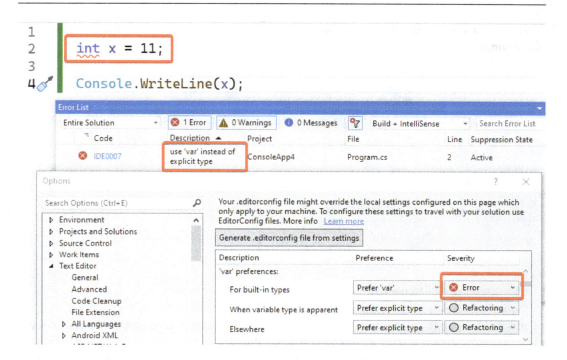

Figure 7.19 – Modifying an option for the .editorconfig file

If we need to specify a configuration file to be applied as part of the solution, even if it is opened on another machine, we can modify the configuration values. Once we have the settings we want to follow throughout the solution, we should click on the **Generate .editorconfig file from settings** button, as shown in *Figure 7.20*:

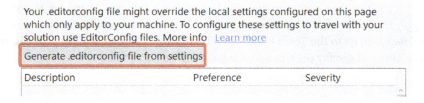

Figure 7.20 – The button to generate a configuration file

This will open a dialogue that asks for the name and path where the configuration file will be saved. In this example, it has been saved as `config.editorconfig`, as shown in *Figure 7.21*:

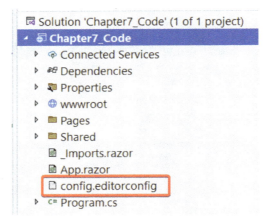

Figure 7.21 – The configuration file has been created

If we proceed to open the file we have created, the text editor will open. Here, we will be able to see the applied configuration in text format and change the preselected parameters quickly. Once again, these changes will accompany the solution so that all source files that are part of the project will have the same code nomenclature.

Now, let's talk about code cleanup profiles, which control the aspects that should be applied when code is cleaned.

Configuring a code cleanup profile

The code cleanup profiles are a configuration in which you can indicate what type of code cleanup you want to apply to your project. There are several ways to access the profile configuration window, but the general way is to go to the **Tools** | **Options** | **Text Editor** | **Code Cleanup** | **Configure Code Cleanup** menu. This will display the window shown in *Figure 7.22*:

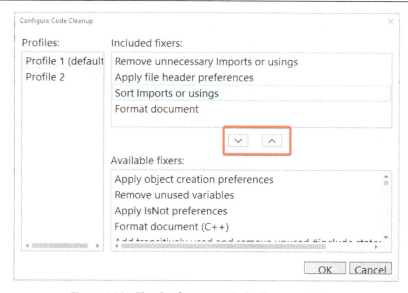

Figure 7.22 – The Configuration Code Cleanup window

As you can see, two cleaning profiles can be configured with different options, with **Profile 1** being the one that will be executed by default. Within each profile, we have two sections:

- The **Included fixers** list contains the specific active actions that we want to apply throughout the project

- The **Available fixers** list contains the actions that are currently disabled, but we could add them to the active actions at any time

Fixers can be enabled or disabled by using the arrow buttons marked in *Figure 7.22*.

Executing code cleanup

Once we have created the code cleanup file and established the fixers that will be applied for the cleanup, we can apply it:

1. Go to the bottom of the editor and press the button with the broom icon, as shown in *Figure 7.23*:

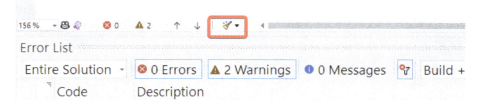

Figure 7.23 – The button to execute code cleanup

As mentioned in the *Configuring a code cleanup profile* subsection, this will apply only the rules that have been configured in the active cleaning profile.

1. To see how a configuration profile might affect your code, select all the fixers for **Profile 1 (default)**:

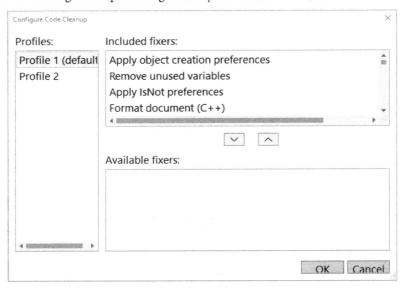

Figure 7.24 – A set of selected test fixers

2. Then, modify the `Chapter7_Code | Program.cs` file by adding a class named `WeatherForecast` to the end of the file, with a complete property and no indentation:

```
public class WeatherForecast
{
private DateTime date;
public DateTime Date
{
get { return date; }
set
{
date = value;
}
}
public int TemperatureC { get; set; }
public int TemperatureF => 32 + (int)(TemperatureC /
            0.5556);
public string? Summary { get; set; }
}
```

Now, when we apply code cleanup, complete properties will be converted to auto-properties and the code will be indented, resulting in cleaner code, as shown in the following code block:

```
public class WeatherForecast
{
    public DateTime Date { get; set; }
    public int TemperatureC { get; set; }
    public int TemperatureF => 32 + (int)(TemperatureC /
                0.5556);
    public string? Summary { get; set; }
}
```

Undoubtedly, code cleanup tools can be of great help to maintain a consistent and clean style, whether we work individually or with a development team.

> **Note**
> It is possible to configure VS to perform a code cleanup every time a file is saved through the **Run Code Cleanup profile on Save** option, which is located in **Tools | Configuration | Text Editor | Code Cleanup**.

Summary

In this chapter, we learned about the different tools that VS has for frontend and backend developers.

We learned how CSS3 snippets can help create cross-browser-compatible styles quickly. Likewise, we saw how hierarchical CSS indentation helps to keep styles readable. We also learned about the color picker, which can help will selecting colors quickly, and we saw how IntelliSense is present when we need to edit CSS files. Finally, we saw how the image editor provides useful tools if we need to make basic edits to our images.

In the case of code analysis, we learned how code quality analysis can help us have safe and reliable code and how it helps us maintain a nomenclature, regardless of whether we are working individually or with a team of developers.

In *Chapter 8*, *Publishing Projects*, you will learn about the most common ways to publish projects on different platforms.

8

Publishing Projects

After finishing a proof of concept or a **minimum viable product** (**MVP**) (which means a demo or pilot project with the main functionalities implemented, as discussed in *Chapter 6, Using Tools for Frontend and Backend Development*), we need to deploy our changes to see how the project works in a real scenario and share the published project with our customers. VS has a set of tools to deploy our projects. We can choose an option to deploy the project in our local environment, but we can also use services in the cloud.

In this chapter, you will learn how to deploy your projects with just a few clicks and VS 2022. The tools described will help you to save time and reduce complexity when you need to deploy.

We will discuss and review the following topics, which are the options to publish projects with VS 2022:

- Publishing web apps
- Publishing .NET MAUI apps
- Publishing desktop apps

Let's dive in and learn all about publishing projects.

Technical requirements

To replicate the concepts in this chapter, the following workloads must be installed using the Visual Studio Installer, as shown in *Chapter 1, Getting Started with Visual Studio 2022*, section *Installing VS 2022*:

- .NET MAUI Multi-platform App UI development
- .NET desktop development
- ASP.NET and web development.

For the *Publishing in Microsoft Azure* subsection, you will need to have an Azure account with credits to complete the deployment.

You can download the projects required for this chapter from the following link: `https://github.com/PacktPublishing/Hands-On-Visual-Studio-2022-Second-Edition/tree/main/Chapter%208`.

Publishing web apps

VS has several ways to access the **Publish** option to configure the deployment of a web project. To follow along with this section, you can open the `Chapter8_Code_Web` project, which you will find in the *Technical requirements* section, or create an ASP.NET Core project.

One of the most common ways of accessing this option is to right-click on the project that we want to publish in the **Solution Explorer** tab (see *Figure 8.1*):

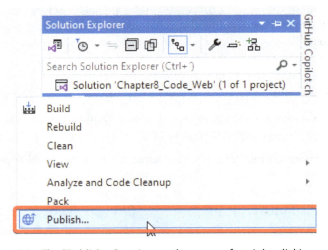

Figure 8.1 – The "Publish…" option on the menu after right-clicking

You can also reach the Publish window through the menu by navigating to **Build | Publish [Project name]**. By default, the main project in the current solution is selected.

Whatever the chosen option, VS will display a modal window to give you the deployment types supported for the main project in the solution or the project selected, as illustrated in the following screenshot:

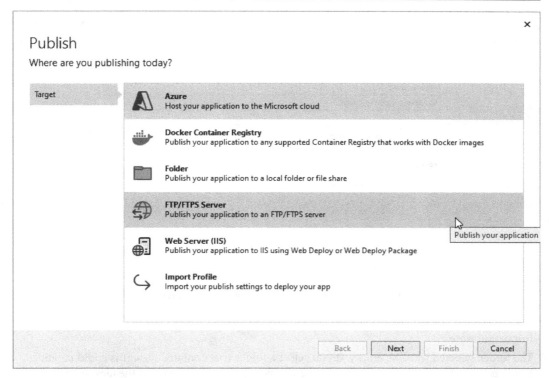

Figure 8.2 – Options to publish a project in VS

Of the options shown in *Figure 8.2*, let us briefly review the most popular ones.

Publishing to a folder

One of the most common options for publishing a project is to use our filesystem and save the site in a folder, including all the resources to use it in a local server, such as IIS, Apache, or NGINX. VS has an option to publish our projects with this approach easily.

To publish to a folder, once you have started the publishing wizard, as shown in *Figure 8.2*, you must select the **Folder** option.

In the next window, we can specify the folder where we want to save the published project using the **Browse…** option. We can use absolute and relative paths. Then, we can finish the configuration by clicking on **Finish**:

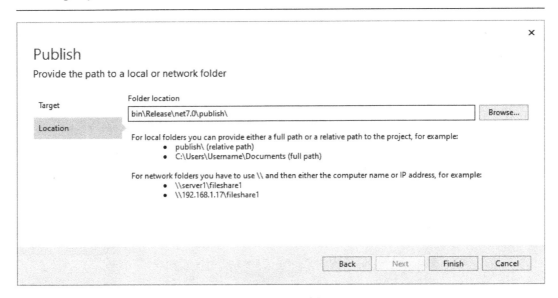

Figure 8.3 – The folder location to publish the project

Important note

The folder selected must be empty. If you select a folder that contains files, this could create conflicts in the files. VS will try to replace the files with the same name using the files generated in the publishing process.

After completing the publishing configuration, VS will generate a file with the `.pubxml` extension that contains the options we chose before in XML format. Now, we can use the **Publish** button to publish the project to the selected folder:

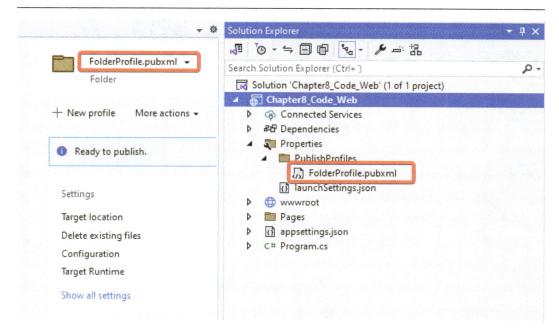

Figure 8.4 – The .pubxml file created with the configuration

Finally, after waiting for some seconds, we will see a **Publish succeeded** confirmation message, which means the process was completed with no issues.

Now let's look at how to publish a web project through IIS.

Publishing to IIS

IIS is the most popular server in the Windows ecosystem. It is included by default in all Windows Server versions, and there is an option to install it on Windows.

To install IIS in Windows 8, 10, 11, or later, you can follow this guide in the *Step 1 – Install IIS and ASP.NET* section of the Microsoft documentation:

`https://docs.microsoft.com/windows/msix/app-installer/web-install-iis`

You can easily deploy a web application in IIS using VS. First, you need to open the **Publish** option that we reviewed in the introduction of the section and select **Web Server (IIS)** (see *Figure 8.2*).

> **Important note**
>
> Although it is valid to have more than one configuration file in a project, to continue with the demonstrations, it is recommended to delete the `.pubxml` file if one has been created, in order not to confuse the steps in the creation of configuration files.

After selecting **Web Server (IIS)**, you can click on **Next** and continue with the process. You will see two options to choose from (see *Figure 8.5*):

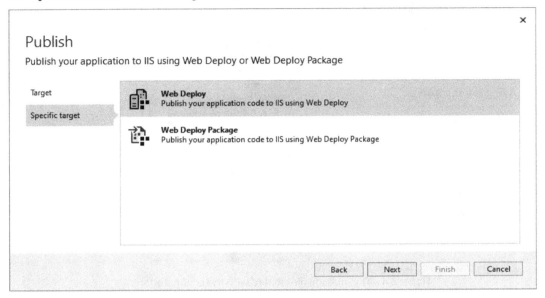

Figure 8.5 – The options to deploy in IIS – Web Deploy and Web Deploy Package

We have two options for deploying in IIS:

- **Web Deploy**: Deploy the folder, including all the files in IIS
- **Web Deploy Package**: Create a `.zip` file, including all the files within the publishing folder

In this case, we will select **Web Deploy** and click on **Next**.

You need to fill in the **Server**, **Site name**, and **Destination URL** fields (see *Figure 8.6*):

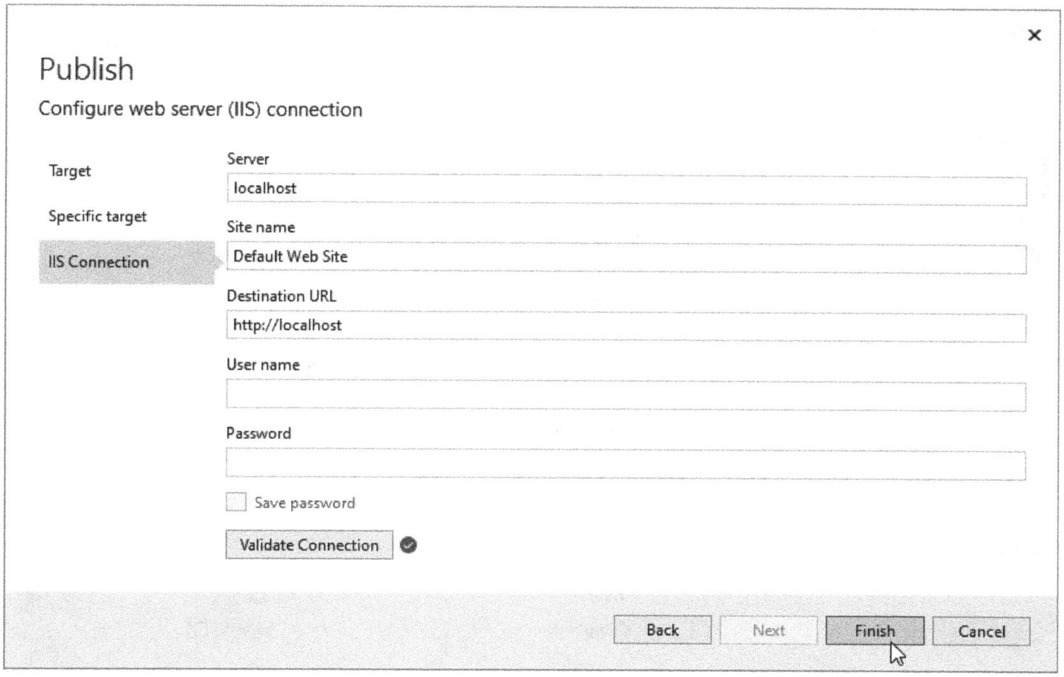

Figure 8.6 – The configuration to publish on a default website in IIS

We will use the default website created in IIS. If you are already using this site with another application, you can set up another site name and/or another destination URL in IIS.

Using the **Validate Connection** button, you can check whether VS can create the site and complete the publication using the setup provided. You can use the **Finish** button to complete the setup.

> **Important note**
> You need to execute VS as admin to give access to IIS; otherwise, you will receive an error.

After completing all the steps, you will get a **Ready to publish** message and some options to edit the configuration. We are ready to use the **Publish** button:

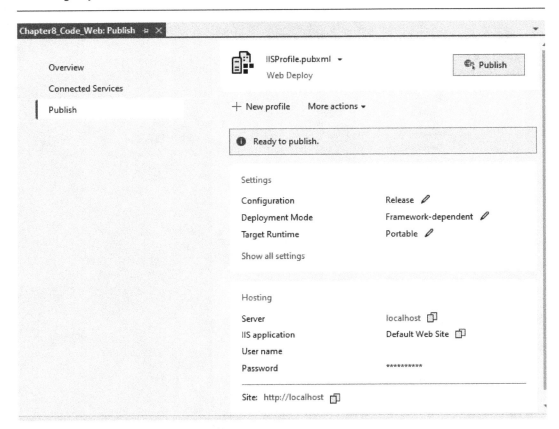

Figure 8.7 – The project ready to publish using IIS

Automatically, VS will open in a new window in the browser, with the URL of the site that we set up in the configuration (see *Figure 8.8*) in the **Destination URL** field. In this case, VS will open `http://localhost:`

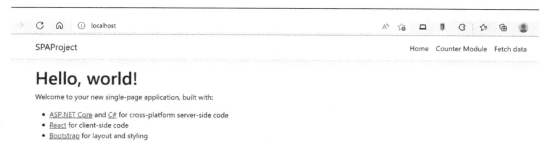

Figure 8.8 – The project running in IIS after publishing it

Now, let's review another way to deploy an application using Azure.

Publishing in Microsoft Azure

Azure is a cloud provider created by Microsoft, and it's one of the most popular among start-ups and .NET developers. Since Azure and VS are supported by the same company and community, there is good integration between both technologies.

To deploy our web project in Azure, we need to select a publish option using the method we reviewed in the *Publishing web apps* section, select **Azure**, and then on the next screen, select **Azure App Service (Windows)**:

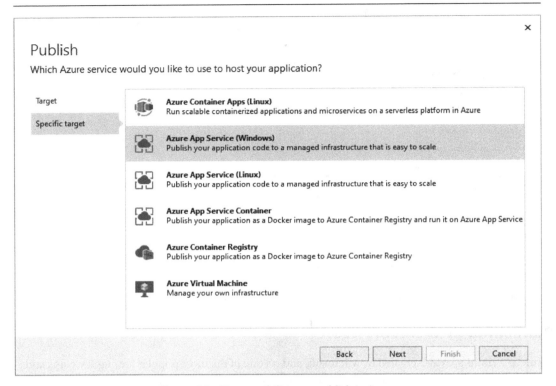

Figure 8.9 – The possibilities to publish in Azure

In *Figure 8.9*, you can see several options for performing a deployment, as well as their description. Remember to delete any `.pubxml` file in the project to create a new configuration to continue with the demonstrations from scratch.

After clicking on **Next**, you will be presented with a new window to perform authentication in Azure and connect with the services and resources in Azure for your account:

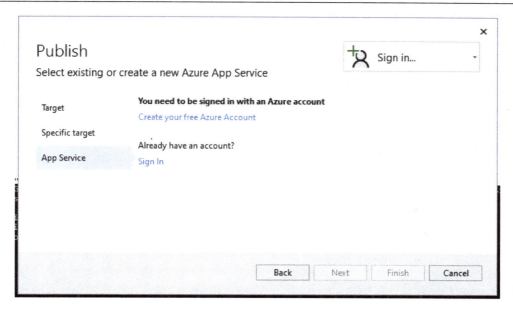

Figure 8.10 – The Azure account sign-in prompt

After performing the authentication, we can select the subscription and create a new web app instance using the green plus button (see *Figure 8.11*):

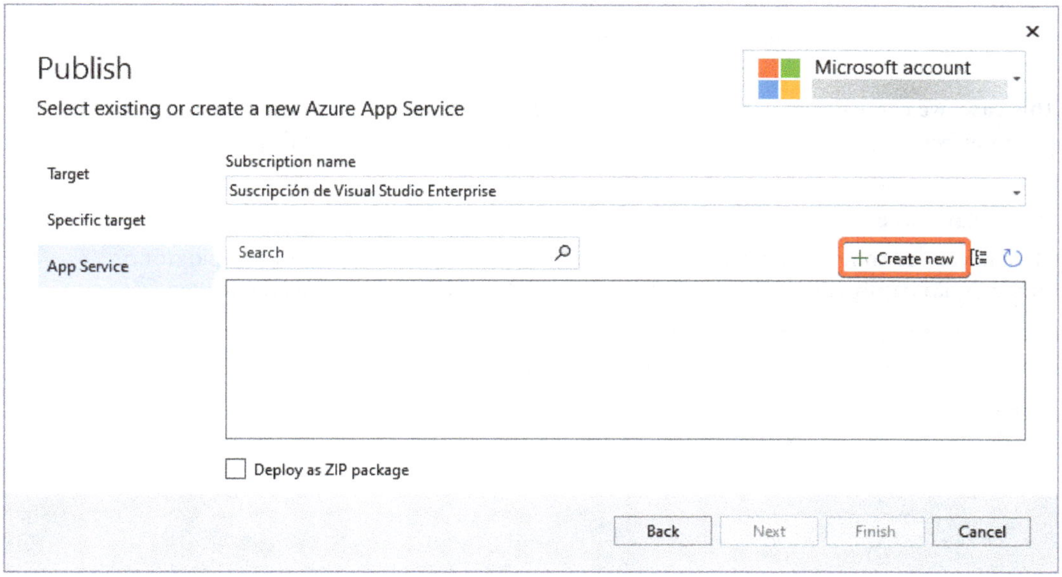

Figure 8.11 – The option to select a subscription and create a new app service

We need to select a hosting plan for our web project:

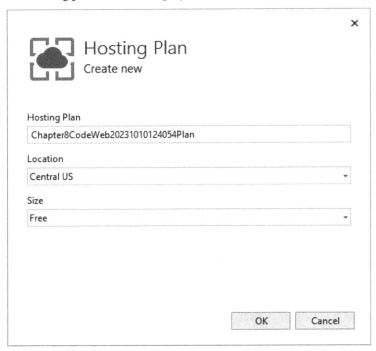

Figure 8.12 – The hosting plan for a new app service

In this case, we can use a free size plan in Azure with limited technical capabilities to run our web app. It's perfect for performing demos and trying this publishing functionality.

> **Important note**
>
> To analyze and compare other plans in Azure, you can navigate to the following link: https://azure.microsoft.com/pricing/details/app-service/windows/.

After creating the hosting plan, we can finish the configuration by choosing a name and resource group for our project (see *Figure 8.13*):

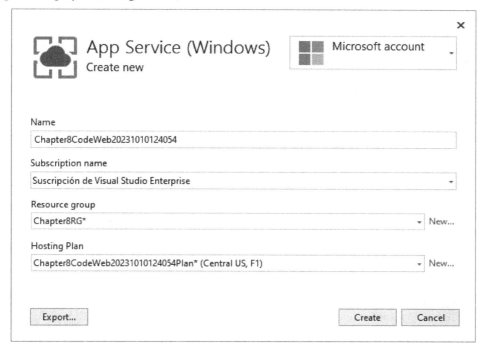

Figure 8.13 – The app service configuration for our web project

In *Figure 8.14*, we can see a preview of the project created. We can complete the configuration using the **Finish** button:

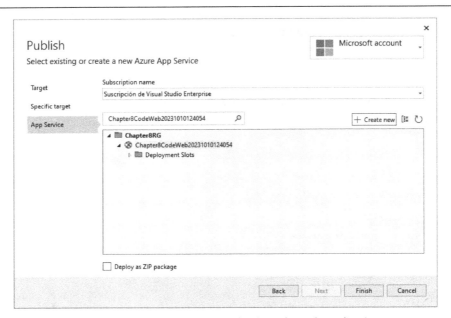

Figure 8.14 – The resource created to host the web application

After creating the resource in Microsoft Azure, you will be shown a message in the same window indicating that a `.pubxml` file with the configuration was created successfully. After closing the window, we need to publish the project using the **Publish** button.

As with *Figure 8.7*, we will see the `.pubxml` file created and the **Publish** button. Click on **Publish** to publish our project in Azure. After publishing the project, VS will open the site using the URL from Azure, which will contain the name of the project followed by the `azurewebsites.net` domain.

You now know how to publish a project in Azure using only the UI in VS and some clicks.

> **More information**
>
> For more details about the integration between VS and Azure, check out this link: `https://docs.microsoft.com/aspnet/core/tutorials/publish-to-azure-webapp-using-vs`.

After reviewing the main options to deploy an application using VS, we conclude this section.

Publishing .NET MAUI apps

.NET MAUI is Microsoft's technology for the creation of multi-platform applications. VS 2022 gives us tools to deploy applications to the different app stores in a simple way, in addition to providing us with ways to test the installation on local computers.

In this section, I will show you how to deploy a .NET MAUI application on the Android platform as it is the most common platform for the deployment of mobile apps.

Publishing an Android app with .NET MAUI

Once you have finished creating your .NET MAUI application, you will probably want to show it to the world through an application store.

We are going to do a test with the project called `Chapter8_Code_MAUI`, the link for which can be found in the *Technical requirements* section. Once you have opened this project, the first step is to assign a **Release** profile to your project, as shown in *Figure 8.15*:

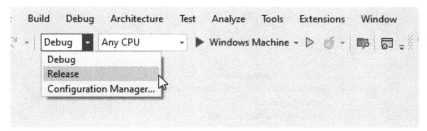

Figure 8.15 – Switching the project to Release mode

Similarly, the platform for which the binary files will be generated must be selected – in our example, the **net8.0-android** platform:

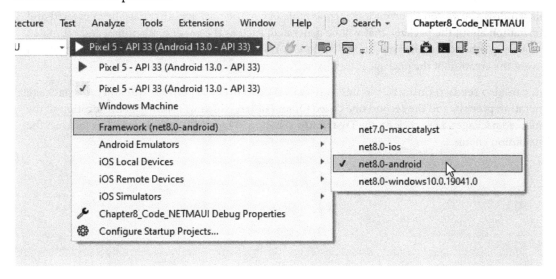

Figure 8.16 – Selecting the deployment platform

Finally, we must right-click on the project and select the **Publish** option, which will display a different window according to the selected platform. In the case of Android, the **Archive Manager** window will open, which will start a checking process to validate that there are no errors. In *Figure 8.17*, you can see the **Archive Manager** window:

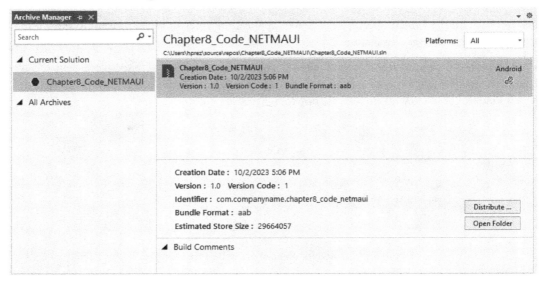

Figure 8.17 – The Archive Manager window

This window shows you information such as a history of the packages that have been created and information about the package that will be generated, such as the version, identifier, and estimated size, among other data. Likewise, in the **Build Comments** section, you will see the errors that will prevent the creation of the package if they exist.

You can also see two buttons, the first one called **Distribute …**, which will start the package generation process, and the second one called **Open Folder**, which will show you the location of the generated packages. Let's click on the **Distribute …** button, which will open the window to select the distribution channel:

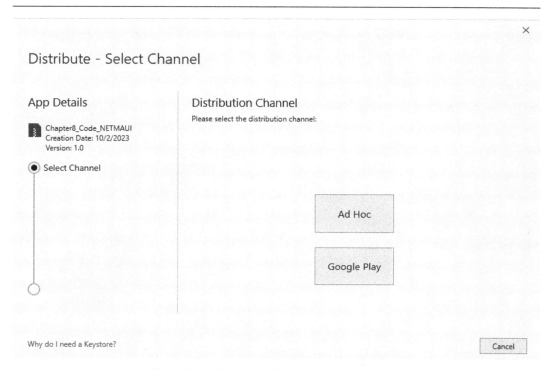

Figure 8.18 – Selection of the distribution channel

The options shown in *Figure 8.18* are detailed as follows:

- **Ad Hoc**: This will initiate the process for generating a local .aab file, which you can distribute and install on different Android devices, either for testing on the device or in case you don't want it to be a public app. The term .aab stands for Android App Bundle and is the publishing format currently used by Android that contains the compiled code and resources of an application. It should be noted that the generated .aab file can also be uploaded to the Play Store manually.

- **Google Play**: This will start the process for publishing an application directly to the Google app store.

> **Note**
>
> To publish an application on Google Play, you must have an Android developer account, which you can obtain at the following link: https://play.google.com/console/signup. It will required you to pay a one-time $25 fee.

We will press the **Ad Hoc** button, which will open the **Distribute - Signing Identity** window. This window is used to specify a digital signature for the generated .aab file, which will verify that the application comes from a trusted source and that it has not been modified after being generated. In this window, press the button with the + symbol to create a new digital signature:

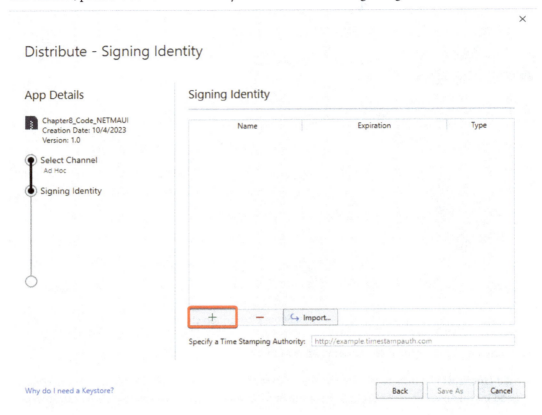

Figure 8.19 – Certificate manager for application signature

After pressing the button with the + symbol, you will be presented with a new window to create an Android keystore, which you can use to create a new certificate that will be used to sign Android applications. You must fill in the keystore information according to information related to you or your organization, as shown:

Figure 8.20 – Window to create an Android keystore

You must enter a password that you will remember because when you use the certificate to sign the packages, you will be asked for the password to complete the process. Next, click on **Create**.

After creating the certificate, you can select it in the list shown in *Figure 8.19*, then click on the **Save As** button, which will open the Windows File Explorer to select a path where the resulting .aab file will be saved. In my case, I have selected the C:\android path, but you can use any folder you want.

After selecting the path where the package will be generated, you will be prompted to enter the password of the generated certificate:

Figure 8.21 – Request for the password of the certificate for the signature of the application.

If everything went well, you will see an icon in the package list indicating that the package was successfully created, and the **Open Distribution** button will be enabled, which, when pressed, will open the path where the .aab file has been generated.

Now that you know the necessary steps to publish an app developed with .NET MAUI in the Google Play Store, let's see how to publish desktop applications with VS 2022.

Publishing desktop apps

ClickOnce is the technology used in VS 2022 to create installers for your desktop applications in a simplified way. This technology allows developers to publish and update their applications from a server or a shared network location so that application users can download, update, and install the app with just a couple of clicks.

For this demonstration, I have created a Widows Presentation Foundation (WPF) application named Chapter8_Code_WPF replicating the game Tic Tac Toe, which you can find in the download link in the *Technical requirements* section.

Once you have opened the project, you can run it, review the source code, or debug it, whatever you want to do. Then, go to the **Solution Explorer** and right-click on the project name (in our case **Chapter8_Code_WPF**) then **Publish…**, which will open a publishing wizard.

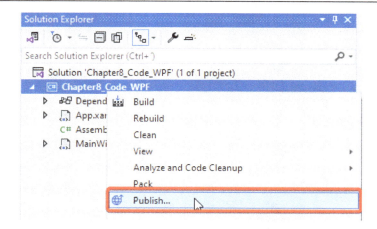

Figure 8.22 – The "Publish…" option to start the ClickOnce wizard

In this wizard, you must select the **ClickOnce** option, which, as mentioned before, is the technology for publishing desktop apps:

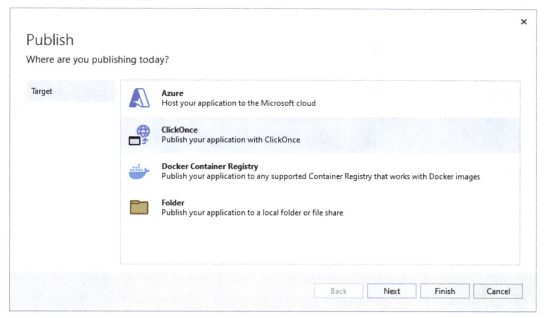

Figure 8.23 – Selecting the "ClickOnce" option

The next window will prompt you for a path, either local or a file share to host the resulting installer files. For demonstration purposes, I have created a folder in the C:\ClickOnce path, which you can also create, or set up a different path. This is the path I will use in the wizard:

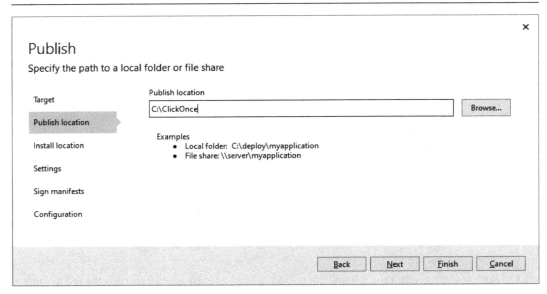

Figure 8.24 – Option to specify the publication location

Subsequently, you will be prompted for the installation location, which will be the path from where users will be able to access the installer. In *Figure 8.25*, you can see three available options:

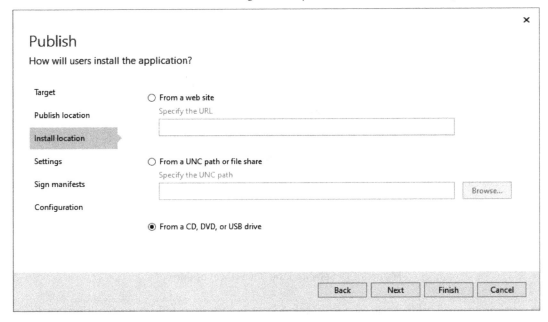

Figure 8.25 – The option to specify how the installation files are to be accessed

Let us take a look at each of these options in detail:

- **From a web site**: With this option, users must access to the installation files through a URL to the application manifest, which contains application information such as name and version, among other data.

- **From a UNC path or file share**: By selecting this option, users will be able to access a file share on a network, which is ideal for enterprise environments where users need to access shared applications within the same network.

- **From a CD, DVD, or USB drive**: With this option, the installer files will be accessible from physical media such as CDs, DVDs, or USB sticks.

In our example, we will select the **From a CD, DVD, or USB drive** option.

The next window we will see is the installer options configuration window (see *Figure 8.26*). Here we will be able to configure which files of the application we want to include, prerequisites to be able to install the application, and options relating to the application, such as its description and extensions association.

Similarly, we can specify whether we want to notify users about possible updates to the application, for which we must specify a valid UNC path or website.

Finally, it is possible to indicate the version of the application and whether we want the version to be increased automatically each time we run the publishing process. In our example, let's leave everything as it is:

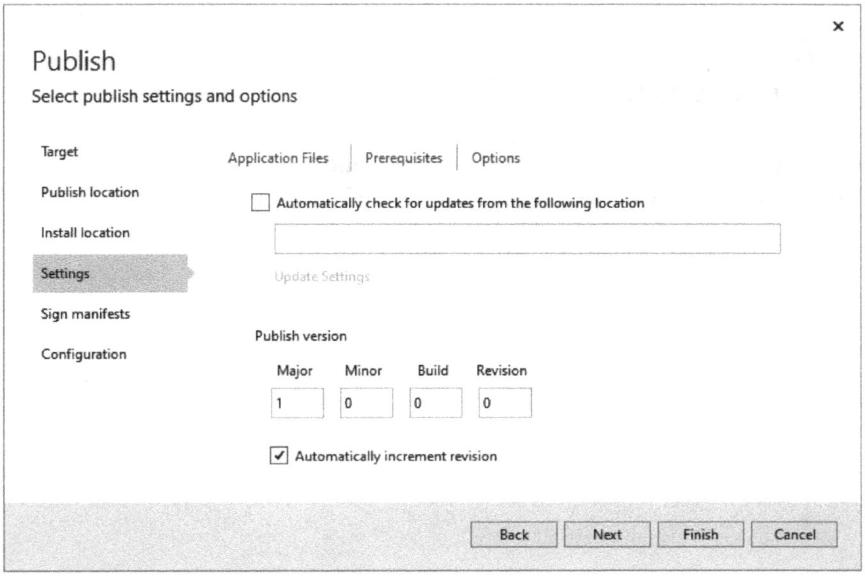

Figure 8.26 – The configuration section of the application

Once we jump to the next window, we will see the options for signing the application manifest. Here, you can select a certificate you already have or create a test certificate, through the **Create test certificate** option:

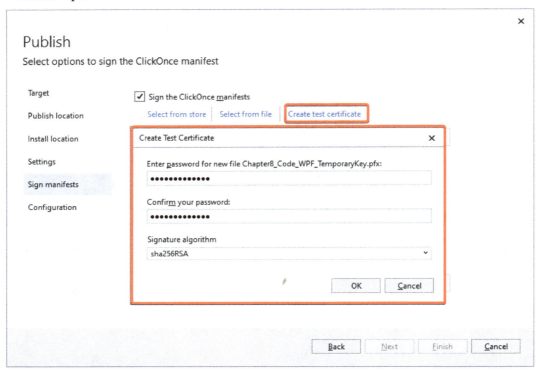

Figure 8.27 – Creating a test certificate for signing the application

To create the test certificate, you should enter a password and select a signing algorithm. After you have created it, the publishing window will be filled with information about the selected certificate:

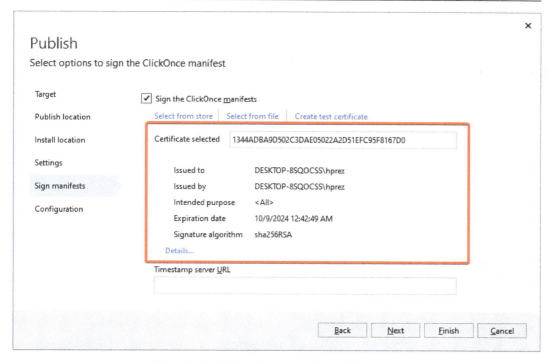

Figure 8.28 – Selected test certificate information

The next section of the wizard will show you the project configuration options, among which we find the following:

- **Configuration**: Allows us to select the project compilation mode. Usually, **Debug** and **Release** modes are available, with **Release** being the recommended option.

- **Target framework**: The version of the framework used to compile and execute the application.

- **Deployment mode**: We have two options:

  - **Framework-dependent**: In this mode, the framework must be installed on the operating system for the application to work. It generates a smaller installer.

  - **Self-contained**: In this mode, a copy of the .NET Framework is added to the resulting assembly, allowing the app to run in environments where the framework is not installed. This causes the installer to be larger but eliminates the need for the framework dependency on the target machine.

- **Target runtime**: Specifies the platform and architecture of the system where the application will be executed.

In this example, we will leave the default options as shown in *Figure 8.29* and click on **Finish**:

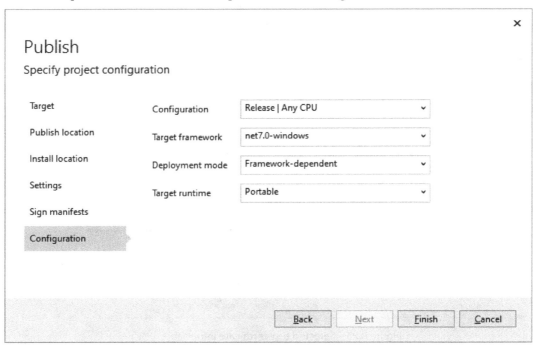

Figure 8.29 – Options for specifying Configuration, Target
framework, Deployment mode, and Target runtime

When you click on **Finish**, a message will appear notifying you about the successful creation of the
.pubxml file.

What this means is that the publication profile with the selected options has been successfully created.
You can close the window by clicking on the **Close** button, which will return you to VS 2022.

With the new profile created, simply click on the **Publish** button to start the installer creation process:

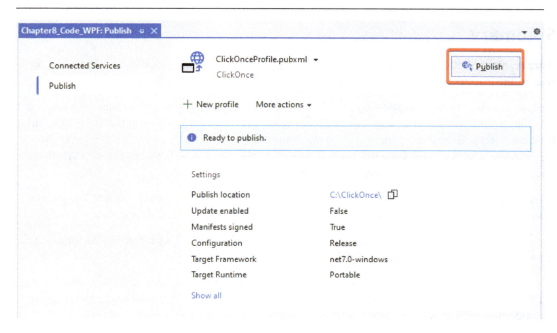

Figure 8.30 – Publication profile successfully created

After publishing the installer, you will be able to find it in the path specified in **Publish location**, so you can run it and test the installation of the application locally:

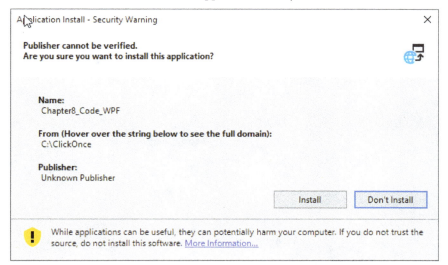

Figure 8.31 – WPF application installer being executed

With this, you have seen how ClickOnce greatly facilitates the publication of desktop application installers.

Summary

After reading this chapter, you have learned about different ways to publish .NET-based projects, for web, multi-platform, and desktop applications.

First, we reviewed some of the options available in VS 2022 for deploying web applications. After reading this chapter, you now know how to use the **Folder** option, deploy your web app in the filesystem, and then use a server to host an application. You also know how to deploy a project in the IIS server using VS. You learned how to deploy a project in Azure using the Azure App Service option in Windows and a publishing configuration using the .pubxml file.

Afterward, you observed the step-by-step process to generate an .aab file from a .NET MAUI project, which you could install on an Android mobile device or publish in the Google app store.

Finally, you saw how to perform a deployment using the ClickOnce tool, which greatly facilitates the generation of installers for desktop applications.

In *Chapter 9, Implementing Git Integration*, we will review all the tools included in VS to connect with Git repositories and GitHub in particular. You will learn how to see the status of your changes using a visual interface and publish your project in a public or private repository easily.

Part 3:
GitHub Integration
and Extensions

In this part, you will learn how to add more functionalities in Visual Studio 2022 using the extensions and how to manage Git repositories.

This part has the following chapters:

Implementing Git Integration

Having a change control platform for the development of a software project is essential for good control of the project. There are many different versioning systems, but **Git** is the most widely used system today, which is why more and more IDEs are including tools for managing repositories based on this technology natively.

This is precisely what happened with VS, which integrates a series of options to allow us to work with Git-based repositories.

In this chapter, you are going to learn how to work with Git repositories based on **GitHub**, which is the most popular repository-hosting platform today.

The main topics we will see in the chapter are as follows:

- Getting started with Git settings
- Creating a Git repository
- Cloning a Git repository
- Fetching, pulling, and pushing Git repositories
- Managing branches
- Viewing changes in repositories

Technical requirements

To follow along with this chapter, you must have installed VS with the workload set from *Chapter 1, Getting Started with Visual Studio 2022*.

Since the projects hosted in the main repository of the previous chapters already have a GitHub configuration, we will create a new project throughout the chapter to perform the practices. Therefore, a GitHub account is required, which can be created at `https://github.com/signup`.

Getting started with Git settings

Starting to work with Git tools is very easy in VS 2022 since they are included as part of the installation itself, so you can install VS and start working on your projects as soon as possible.

To access the management of code projects hosted on GitHub, you must first sign in with a Microsoft account, as explained in *Chapter 1, Getting Started with Visual Studio 2022*. Once logged in, click on the account profile icon and select the **Account settings** option, as shown in *Figure 9.1*:

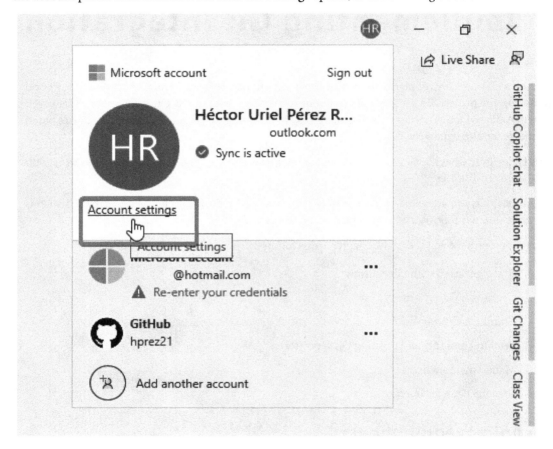

Figure 9.1 – Accessing the account settings

This will open an account customization window, where we can add a GitHub account, via the **+Add** button, as shown in *Figure 9.2*:

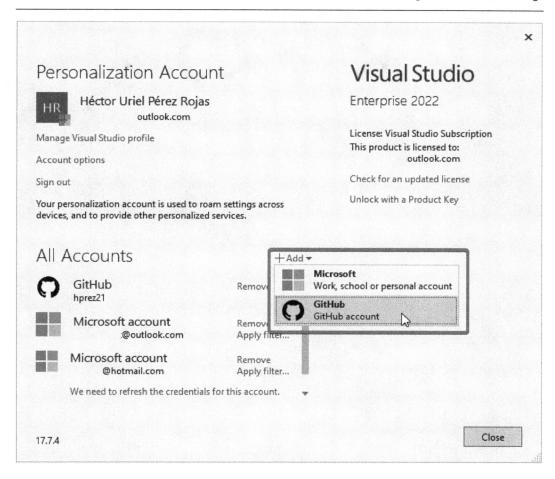

Figure 9.2 – The "+Add" button for adding a GitHub account

Once we press the button, we will be redirected to the GitHub authentication portal, where we can log in with an existing GitHub account or create an account if we don't have one. After successful authentication, we will be asked to authorize VS to interact with GitHub services, so it is essential to press the button that says **Authorize github**:

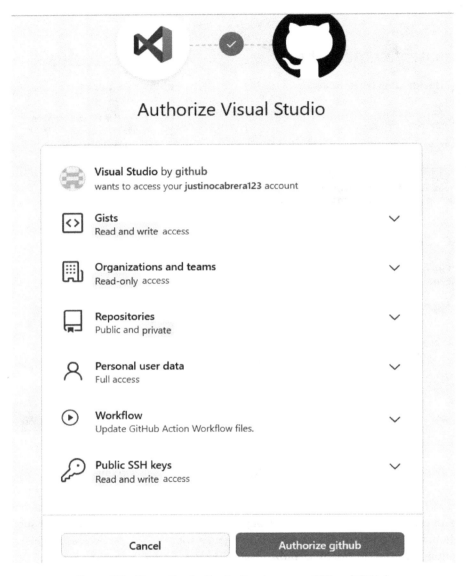

Figure 9.3 – The authorization button to connect VS and GitHub

After this step, the GitHub account will have been added as part of the accounts associated with the main VS account. Now, we can start working with the repositories of the GitHub account, as shown in *Figure 9.4*:

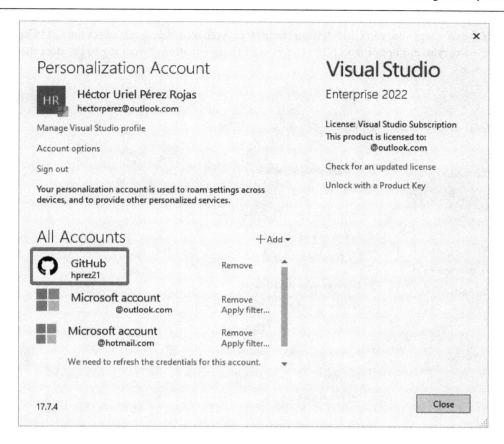

Figure 9.4 – GitHub account added to the account listing

Now that we have associated a GitHub account with VS, let's learn how to create Git repositories in GitHub.

> **Note**
>
> Although we will work with GitHub in this chapter, VS supports working with other providers, such as Azure DevOps, GitLab, Bitbucket, and so on, through the installation of additional extensions.

Creating a Git repository

Creating a repository in GitHub from a VS project is very easy to do. In this section, you will have to test your knowledge by creating a new **ASP.NET Core Empty** project, as discussed in the *Templates for ASP.NET Core web applications* section of *Chapter 2, Creating Projects and Templates*. You can name this project GitDemo.

To create the new repository in GitHub from the project you have created, just select the **Git | Create a Git Repository** menu option and fill in the repository information, according to the data shown in *Figure 9.5*:

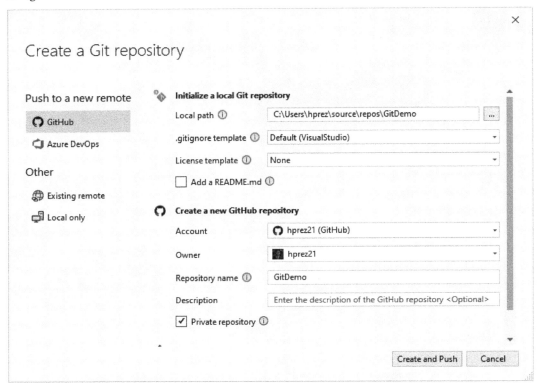

Figure 9.5 – Filling in the information for a new repository on GitHub

Let's briefly look at each option:

- **Local path**: This sets the path on the local machine where the source code is hosted (usually the path where you created the project).

- **.gitignore template**: This allows you to select a template that establishes a set of files that will not be uploaded to the repository—for example, files that are generated after a compilation and can be regenerated on a different computer are selected by the selected template.

- **License template**: This allows you to choose a license for the repository code, which indicates what users can and cannot do with the project code. Some of the most widely used licenses are MIT, Apache-2.0, GPL-3.0, among many others.

- **Add a README**: This allows you to add a README file that describes the purpose of the repository using Markdown style.

- **Account**: This allows you to select the GitHub account to which the repository will be published. It is possible to associate an account from here if one has not been associated before.

- **Owner**: This allows you to set which GitHub account will be the owner of the repository if the account belongs to several work teams.

- **Repository name**: This allows you to set the name of the repository, although the project's own name is normally used. This will affect the URL of the final repository.

- **Description**: This allows you to enter a description of the repository created.

- **Private repository**: This allows you to set whether the repository will be public or private.

Once these values are set, just click on the **Create and Push** button to start the repository creation process, the result of which can be seen in *Figure 9.6*. It is important that you follow this step in your GitHub account so that you can complete all of the activities in the remaining sections of this chapter.

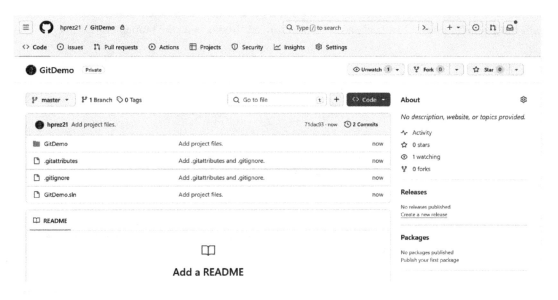

Figure 9.6 – The project created on GitHub

There are several signs that can help you identify whether a project belongs to a Git repository, as shown in *Figure 9.7*. One sign is lock icons on the left side of source files, which indicate that they are the original repository files:

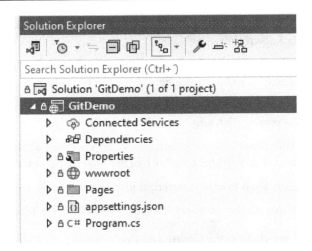

Figure 9.7 – Signs to identify a versioned project with Git

Another sign that a project belongs to a Git repository is a branch icon at the bottom (as shown in *Figure 9.8*), which indicates the branch we are working on. In our case, the branch is `master`:

Figure 9.8 – Branch selection enabled with the master branch by default

Now that we know how to create a new repository, let's see how to clone a repository to our local machine.

Cloning a Git repository

You may want to clone an existing repository and not start from scratch because you need to work in a team. Alternatively, you may simply have come across a repository that caught your attention while browsing the GitHub site.

The easiest way to clone a repository is from the initial window of VS, which can be reached by either starting a new instance of VS, closing an open project, or going to **File | Start Window**. In this window, the first option is the **Clone a repository** button:

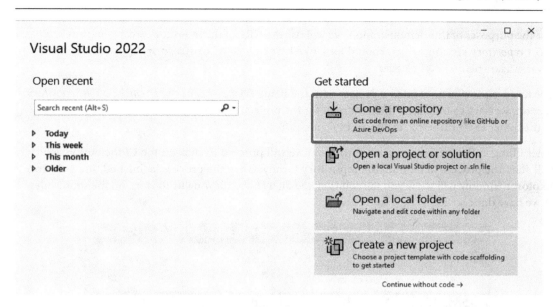

Figure 9.9 – The "Clone a repository" option from the startup window

Once we press this button, a new window (shown in *Figure 9.10*) will open, asking us to indicate the URL of the remote repository and the local path where the source code files will be stored:

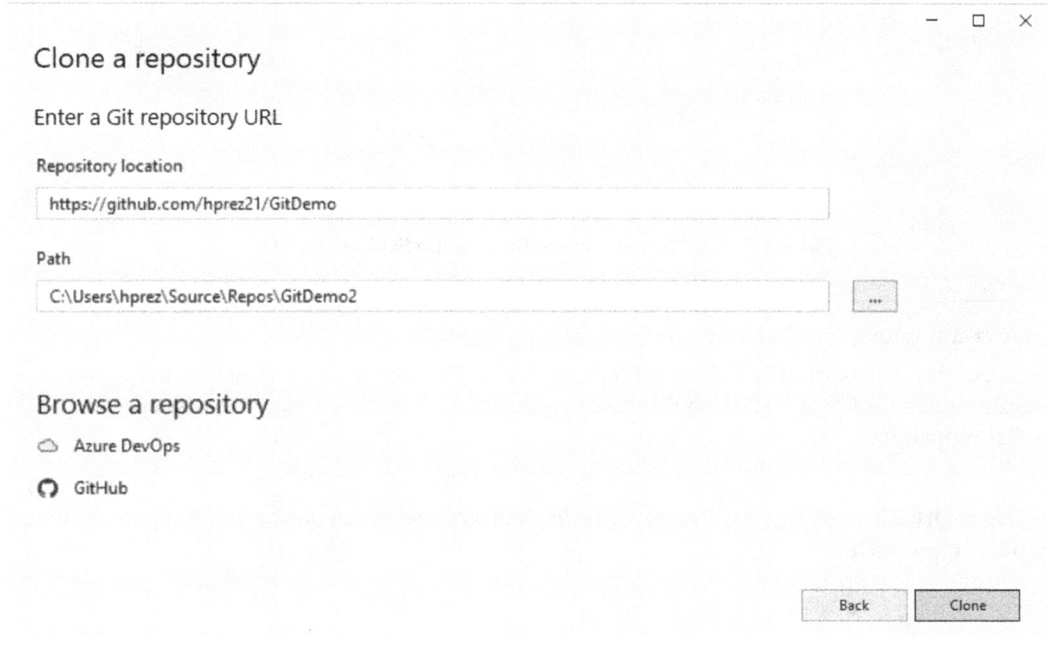

Clone a repository

Enter a Git repository URL

Repository location

https://github.com/hprez21/GitDemo

Path

C:\Users\hprez\Source\Repos\GitDemo2

Browse a repository

☁ Azure DevOps

◯ GitHub

Back   Clone

Figure 9.10 – The window for cloning a repository

For the purposes of this demonstration, we will use the URL of the repository created in the *Creating a Git repository* section, which should look like `https://github.com/{your-github-username-here}/GitDemo`.

Because we previously created a project with the name `GitDemo` in the *Creating a Git repository* section, we have to indicate a different path for the project. For simplicity, we will change the name of the folder to `GitDemo2`, as shown in *Figure 9.10*.

After filling in the information in the window, we will proceed to click on the **Clone** button, which will start the process of cloning the repository locally. Once the process is finished, the **Solution Explorer** window will show you the solution and all its files, allowing us to work on the source files, as we have done so far:

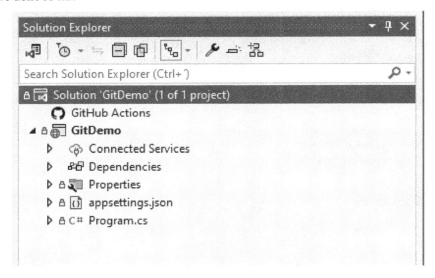

Figure 9.11 – The folder view on the Solution Explorer window

> **Important note**
> It is possible to change the default folder view to the solution view by selecting the option located at **Git | Settings | Git Global Settings | Automatically load the solution when opening a Git repository**.

Now that we have learned how to clone repositories, let's see how we can perform pushing and pulling actions on repositories.

Fetching, pulling, and pushing Git repositories

The most important commands when working with Git repositories have to do with fetching, pulling, and pushing operations. There are two main ways to execute these operations:

- The first way is by accessing them through the **Git** menu, as shown in *Figure 9.12*:

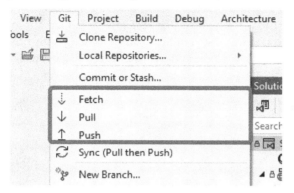

Figure 9.12 – Accessing the Fetch, Pull, and Push operations from the menu

- The second way is to enable the **Git Changes** window, which you can open through the **View | Git Changes** menu, as shown in *Figure 9.13*:

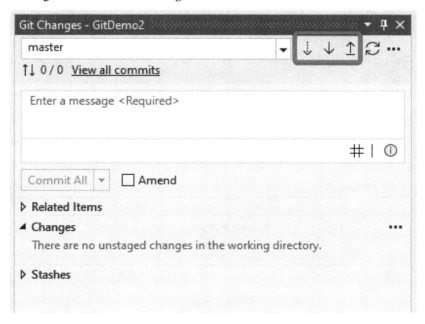

Figure 9.13 – Accessing the Fetch, Pull, and Push operations from the Git Changes window

At this point, you may be wondering what each of these operations is for. So, let's learn about them briefly.

Fetching repositories

The fetch operation allows you to check whether there are remote commits that should be incorporated into the local repository.

To run this example, go to the GitHub portal, log in if you are not already logged in, and open the repository called GitDemo that we created in the *Creating a Git repository* section. Once you are in the repository, go to the Program.cs file and click on the pencil icon, which will allow you to edit the file, as shown in *Figure 9.14*:

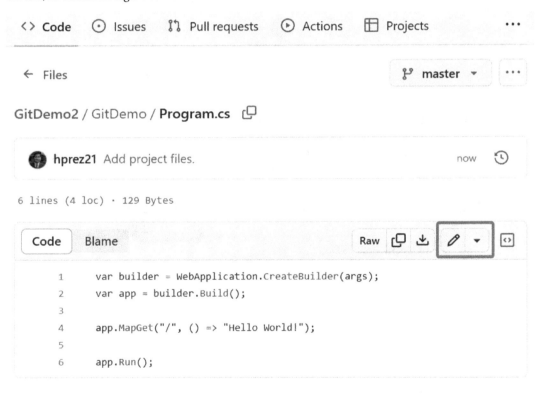

Figure 9.14 – The button to edit a repository file

We will make a very simple change by adding a pair of exclamation marks at the end of the string on line 4 of the code in *Figure 9.14*:

```
app.MapGet("/", () => "Hello World!!!");
```

Once this change has been made, click on the green button to commit the changes:

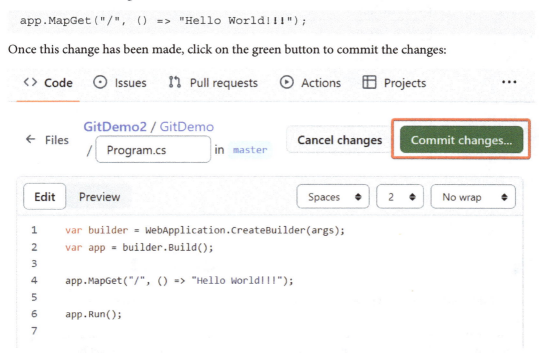

Figure 9.15 – The button to commit changes on the repository

After pressing **Commit changes...**, a new window will be displayed asking for information about the change. The information requested includes a message to be placed to identify the change made and a text box to add a longer description if needed. Finally, we can select if we want to commit directly to the root branch or if we want to create a new branch. For our example, we will leave the default data and click on **Commit changes**:

Figure 9.16 – Window for adding information about the changes made in the commit

If we now go to VS and click on the **Fetch** button, as shown in *Figure 9.17*, we can see a **0 outgoing / 1 incoming** message, indicating that there is a change in the repository that has not yet been applied to the local project:

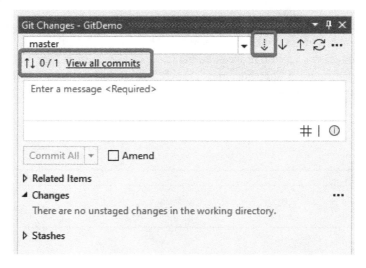

Figure 9.17 – The button for fetching from the repository

Additionally, if we click on the **0 outgoing / 1 incoming** message, a new window will open, which will show us the version history of the project, indicating the changes that we have not applied locally:

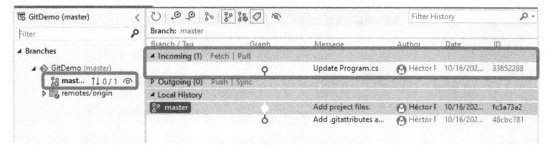

Figure 9.18 – The window showing the pending changes to be applied to the local repository

It should be noted that the fetch operation does not make any changes locally or in the remote repository, which the pull and push operations do. We will discuss these operations in the upcoming section.

Pulling repositories

A pull operation refers to the act of downloading the latest changes from the repository to our local project. In the *Fetching repositories* section, we saw that we have a pending change to apply. So, we will proceed to click on the **Pull** button, as shown in *Figure 9.19*, to apply it:

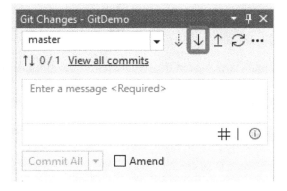

Figure 9.19 – The button for pulling changes

Once the changes have been downloaded, a message will appear, indicating which commit has been applied to the current project, as shown in *Figure 9.20*:

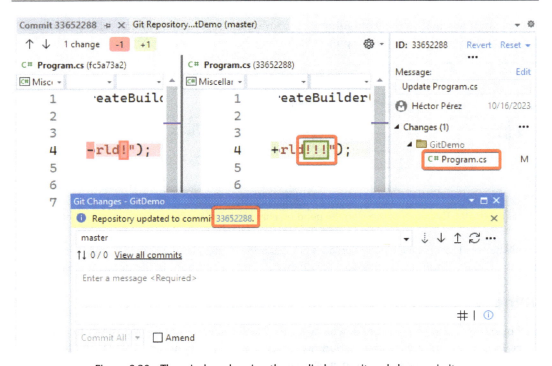

Figure 9.20 – The window showing the applied commit and changes in it

If we click on the name of the commit, a window will appear showing the changes that have been applied to the project files, as shown in *Figure 9.20*.

In this section, although we made a change in the remote repository from the main GitHub site, the most common way to do it is in VS itself. That is why, in the next section, we will analyze how to perform push operations on the repositories.

Pushing to repositories

A push operation refers to uploading local changes to a repository. To demonstrate how this operation works, let's open the `Program.cs` file and modify line 4 of the code in *Figure 9.14*, changing the `Hello World` string to `Hello Git`:

```
app.MapGet("/", () => "Hello Git!!!");
```

Immediately, you will see that the file icon in the **Solution Explorer** changes to a red checkmark, indicating that there has been a change in the local repository, which we can upload to the repository on GitHub:

Figure 9.21 – The checkmark indicating that the original file has been modified

Similarly, in the **Git Changes** window, you will see a list of those files that have changes and can be uploaded to the repository, as shown in *Figure 9.22*:

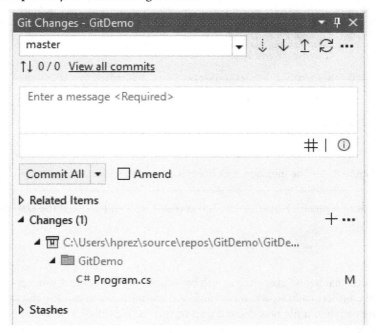

Figure 9.22 – The list of files with changes that can be uploaded to the remote repository

In the preceding figure, we can also see a button with the **Commit All** legend. This button is used to create a commit locally without affecting the remote repository. This is a drop-down button that contains more options we can use, such as the **Commit All and Push** option. This option will push the changes to the remote repository.

For our demonstration, we will add the Modified Program.cs message, as shown in *Figure 9.23*, and click on the **Commit All and Push** option to apply the changes on the server:

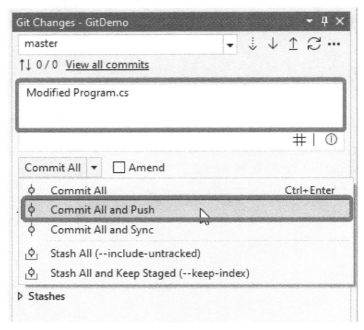

Figure 9.23 – The message added and the Commit All and Push option

If you go to the online repository in your GitHub account, you will see that the changes have been applied successfully.

> **Important note**
> It is recommended that before executing a push operation, you always perform a pull operation to avoid merge conflicts (that is, when two different commits try to modify the same line or section of code or when a file was deleted and tries to be modified in another commit) as much as possible. You can use the **Sync (Pull then Push)** button to perform both actions one after another.

Now that we have analyzed how to execute the most common operations in GitHub, let's see how to work with branches in VS.

Managing branches

So far, we have been working with the main branch of our project called `master`. Imagine this branch as a timeline, where each event is performed by a commit. This is very useful when there is some conflict and you need to go back to a previous version, undoing the changes of a specific commit.

However, if you are working in a team, it is common that you will need to add a functionality in some kind of sandbox before merging this functionality into the master branch. It is in this sort of scenario that Git branches will help us, allowing us to create a new project branch from an existing repository branch and work on it without affecting the functionality of the main repository.

To create a new branch, just go to the **Git | New Branch**. This will open a new window that asks for the branch name—the branch on which the new branch will be based— and a checkbox labeled **Checkout branch**, which, if checked, will transition to the new branch once it is created.

For this demonstration, let's use `branch01` as the name of the branch, which will be based on the `master` branch, and leave the checkbox selected, as shown in *Figure 9.24*:

Figure 9.24 – Creating a new branch

Once the new branch has been created, we can apply as many changes as we need. In our example, we will open the `Program.cs` file and modify line 4 of the code in *Figure 9.14* again as follows:

```
app.MapGet("/", () => "Hello Git!!! - branch01");
```

Once this change has been made, we will push the code as described in the *Pushing to repositories* section, making sure to apply the change to `branch01`, as shown in *Figure 9.25*:

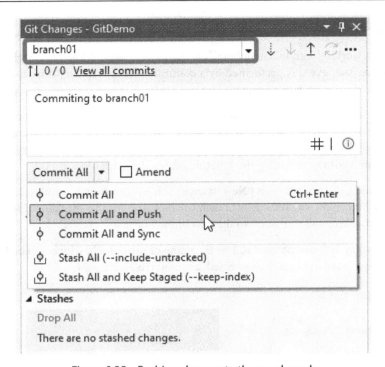

Figure 9.25 – Pushing changes to the new branch

This allows us to add functionality on an isolated remote repository without affecting the operation of the master branch.

Once you have your code tested, you will probably want to integrate it back into the master branch. This can be done from the **View | Git Repository** window. In this window, we have a section called **Branches** where we will see the list of the different branches in our project. Just right-click on the master branch and click on the **Checkout** option to switch to the master branch, which is where we will add the branch01 changes. Then, right-click on the branch01 branch and select the **Merge 'branch01' into 'master'** option:

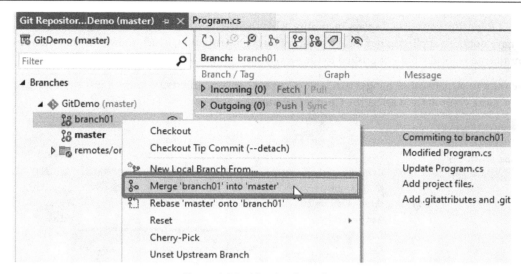

Figure 9.26 – Merging branches

After clicking on the **Merge 'branch01' into 'master'** option, a window will open to confirm the integration. This will cause the `branch01` branch to be merged with the `master` branch, which means that the new functionality will be integrated into the project.

Now that we have seen how to create branches and merge them, it is time to see how VS allows us to visualize the changes while editing the source code.

Viewing changes in repositories

There are several ways in which VS helps us to visualize the changes in the repositories. The first one is through the **Git Repository** window (which, in case you accidentally closed it, can be found in the **View** menu), as shown in *Figure 9.27*:

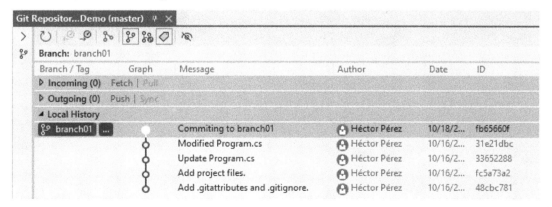

Figure 9.27 – The Git Repository window

The **Git Repository** window displays **Incoming**, which allows us to visualize whether there are versions that have not been applied in our local repository through, **Outgoing**, which shows whether there are commits made locally but not pushed to the server, and also **Local History**.

If we want to visualize the changes that have occurred between different commits, just right-click on two or more commits and select the **Compare Commits…** option:

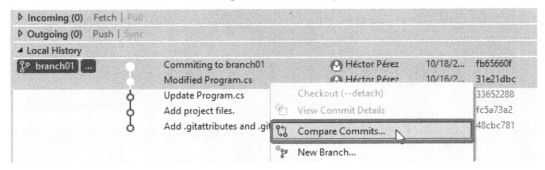

Figure 9.28 – The "Compare Commits…" option

This will display a new window with the changes that occurred between the different source files.

Another way to view changes to a single file is to right-click on a file in the **Solution Explorer** and then select the **Git | View History** option, which will open the changes window for the selected file only, as shown in *Figure 9.29*:

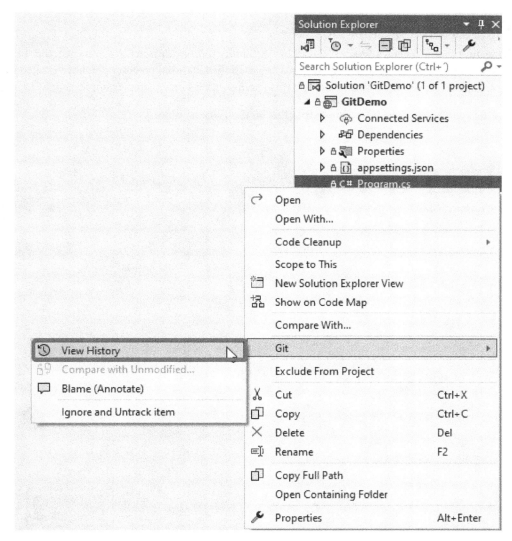

Figure 9.29 – The option to view the history of changes to a specific file

Finally, we already discussed the *CodeLens* functionality in *Chapter 5, Coding Efficiently with AI and Code Views*, which also contains a function that lets you view changes to the members of a class, such as the history of changes and who made modifications to a member.

Now, it's time to talk about a feature that will allow you to reference GitHub issues directly when you make a commit.

Linking GitHub issues

A new feature that has been added to VS 2022 is the ability to link commits to a specific issue. Let's see how to do it in a practical way. For this practice, go to the GitHub repository you have created in the *Creating a Git repository* section. Let's test this feature with a step-by-step example:

1. Once you have opened the repository, click on the **Issues** option:

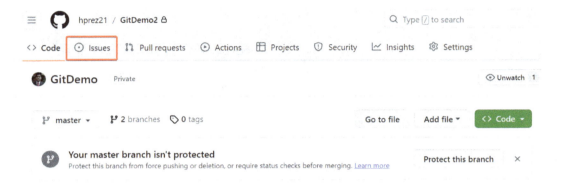

Figure 9.30 – The "Issues" section of the repository

2. Next, press the button on the right side labeled **New issue**:

Figure 9.31 – Button to add a new issue

3. After clicking on the **New issue** button, we will be redirected to a new window, as shown in *Figure 9.32*, where we will raise an issue to indicate that there is a misspelled text. For it I will set the Title to: `Typographical Error in Root Endpoint Response`. In the **Extended description** text box, I will add the following comment in markdown format:

```
### Description:
Typographical error in the greeting message returned by the root
endpoint. It reads "Hello Git!!!" instead of "Hello Git!".
```

```
### Suggested Fix:
Change
```csharp
app.MapGet("/", () => "Hello Git!!!");
```

to
```csharp
app.MapGet("/", () => "Hello Git!");
```
```

4. Finally, press the **Submit new issue** button to create the new issue in the repository, as seen in *Figure 9.32.*:

Add a title

Typographical Error in Root Endpoint Response

Add a description

| Write | Preview | | | H | B | I | ⌵≣ | <> | 𝒫 | ⌵≣ | ≣ | ⌵≣ | 𝒪 | @ | ☐ | ↰ | ☑ |

```
### Description:
Typographical error in the greeting message returned by the root endpoint. It reads "Hello Git!!!"
instead of "Hello Git!".
### Suggested Fix:
Change
```csharp
app.MapGet("/", () => "Hello Git!!!");
```

to
```csharp
app.MapGet("/", () => "Hello Git!");
```
```

Markdown is supported   **Paste, drop, or click to add files**

Submit new issue

Figure 9.32 – Window for adding the description of an issue

5. The next step is to go back to VS 2022 and open the `GitDemo` project in case you don't have it open. Then, open the **Git Changes** window and run a **Sync** operation to synchronize all the changes between the local project and the remote project:

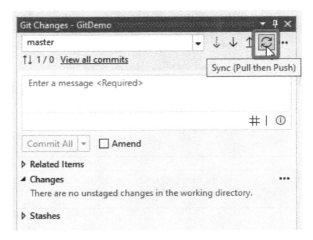

Figure 9.33 – Sync button for executing a Pull operation first and then a Push operation

6. Now, open the `Program.cs` file and replace the highlighted text in the following code line:

```
app.MapGet("/", () => "Hello Git!!! - branch01");
```

to this:

```
app.MapGet("/", () => "Hello Git!");
```

7. Finally, go back to the **Git Changes** window and start typing `Text fixed #`. By writing the # symbol, the existing issues in the project will be displayed, including the one we created at the beginning of this section. Select it from the list and perform a **Commit all and Push** operation.

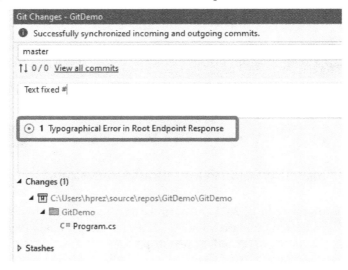

Figure 9.34 – Referencing an issue from the GitHub repository using the # symbol

8. Now, if you go back to the issue on GitHub you will see that it has been marked as **Closed** due to the fix we sent from VS to resolve this specific issue.

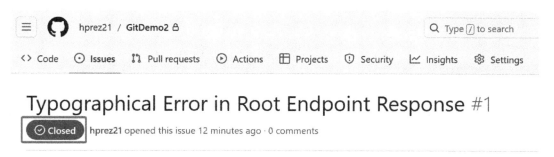

Figure 9.35 – Issue marked as Closed because it has been referenced through a commit

Undoubtedly, being able to link a commit to an issue represents a shortcut when solving problems in a project hosted on GitHub.

Summary

In this chapter, we have learned how VS integrates tools so that we can easily manage projects using Git and GitHub.

Knowing how to work with Git-based projects is indispensable for all developers who want to manage their projects in a more controlled way, while having an overview of a project structure in previous versions.

Likewise, if you work with other developers, you will be able to share project tasks, test them independently, and merge them when the code is reliable in order to not damage the source code that has already been tested.

That is why we have learned how to set up a GitHub account in VS, how to create and clone repositories, how to perform fetch, pull, and push operations, how to manage branches in our projects, how to visualize changes in repositories, and, finally, how to reference repository issues with a commit.

In *Chapter 10, Sharing Code with Live Share*, you will learn how to work collaboratively with a development team on the same project in real time thanks to the use of Live Share.

10

Sharing Code with Live Share

Collaboration tools are a new way to work remotely and make it easier for a group to achieve one common goal. It's amazing to see how we can edit the same document, record, or share resources in real time with other users. If we want to collaborate on coding or perform **pair programming**, which is a technique where two programmers work together on the same code, we need to be working in the same place, on the same machine, or use a tool to make a video call to perform these activities.

VS Live Share, or just **Live Share**, is a new tool included in VS by default that helps us to share our code with other programmers that use VS and VS Code.

We will review the following topics in this chapter:

- Understanding Live Share
- Using Live Share
- Performing live editing
- Sharing a terminal with other collaborators

First, we need to understand Live Share in general and know how to find it.

Technical requirements

To use Live Share in VS 2022, you must have previously installed VS 2022 with the web development workload, as shown in *Chapter 1, Getting Started with Visual Studio 2022*.

To follow the demonstrations in this chapter, you can use any project you have created, or use the GitDemo project found at the following link: `https://github.com/hprez21/GitDemo2`.

Understanding Live Share

Live Share is a real-time collaboration tool for programmers who use VS and VS Code. Live Share was launched as an extension for VS 2017, with some trial features available at that time. In VS 2022, it is included by default and contains all the features.

Live Share has tools to edit, debug, share terminals, and execute projects for remote developers. We don't need to clone the repository or install additional extensions to see code and interact with it.

To complement the information provided in this chapter and the functionalities that you will review in the *Using Live Share* section, you can read the documentation at `https://docs.microsoft.com/visualstudio/liveshare/`.

These kinds of collaboration tools are not new. Other IDEs have extensions and components to share code and work in real time. You can try out the following tools and compare them with Live Share:

- **Duckly** (`https://duckly.com/`): A real-time collaboration tool with video calls and other features compatible with many IDEs.

- **CodeTogether** (`https://www.codetogether.com/`): A tool to create shared coding sessions that supports VS Code, IntelliJ IDEA, and Eclipse.

> **Important note**
> There are no other free collaboration tools for VS. Live Share is supported by Microsoft and the community.

In this section, we provided a brief overview of Live share, its history, and alternative tools that perform similar tasks. Now, let's start to use Live Share.

Using Live Share

To start using Live Share, we can go to the icon located at the top right of the main window in VS, as shown in *Figure 10.1*:

Figure 10.1 – The Live Share button in the main window

After clicking on this icon, if you have not previously logged in with a Microsoft or GitHub account, you will see a new window, where you must select the account to use to create a new live share session:

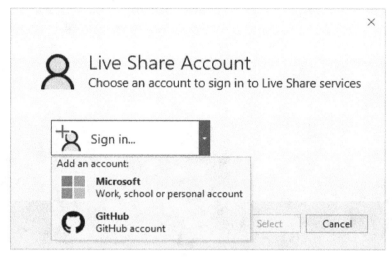

Figure 10.2 – The Live Share account window

You can select the account that you want to use and then click on **Select**. Then, you will get a confirmation message, and the link to share the session is automatically added to the clipboard. The Live Share button displays **Sharing**, which confirms that the session is live (see *Figure 10.3*):

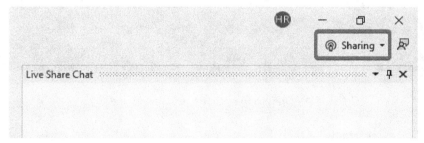

Figure 10.3 – VS with a live session in Live Share

You can share the invitation link with other developers, coworkers, or friends to try this tool. This is an example of a link generated by Live Share: `https://prod.liveshare.vsengsaas.visualstudio.com/join?7E72234CE1703CF92015D01564C560706AE1`.

When someone opens the link in a browser, they will see **Visual Studio Code for the web**. This is a version of the VS Code editor that runs as a web page, with no dependencies on the operating system or additional requirements. To read more information, visit this link: `https://code.visualstudio.com/docs/editor/vscode-web`. Three options will be shown that we can choose to edit and navigate in the code (see *Figure 10.4*):

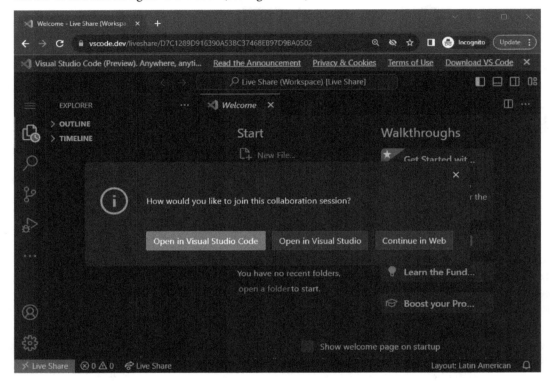

Figure 10.4 – Options to open the session with the link

Let's review these three options further:

- **Open in Visual Studio Code**: This option will open VS Code on your computer, and you will be able to edit the code there

- **Open in Visual Studio**: This option will execute VS on your local machine, and you will be able to edit the code there

- **Continue in Web**: With this option, you will continue using VS Code in the browser

For this demonstration, select the **Continue in Web** option to join in a session using the browser.

You will then be prompted to log in or continue anonymously to join the session. This provides an additional security step:

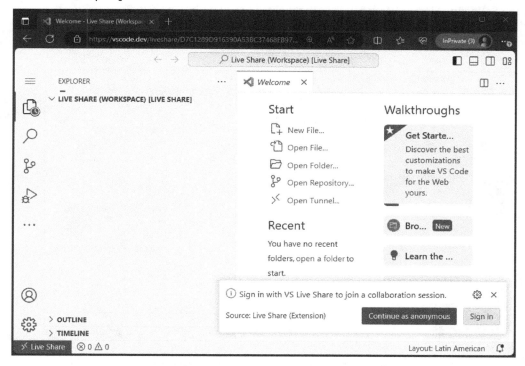

Figure 10.5 – A session opened in the browser

We can use **Continue as anonymous**, which will prompt you for an alias to use during the session.

The user that created the session will see a new notification message, where you can see the collaborators that are trying to join the session. You can accept the new collaborator and continue with the session (see *Figure 10.6*):

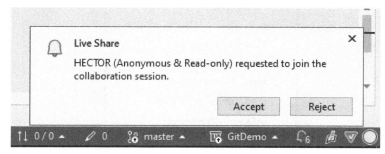

Figure 10.6 – A notification from Live Share to accept a new collaborator

After accepting the guest user, Live Share will share the code and show the changes in real time. In this case, the guest user can see the moves in code performed by *Héctor Pérez*, who created the session. See the **Following Héctor Pérez** message, highlighted in yellow, in *Figure 10.7*:

Figure 10.7 – The option to open the session in VS

In VS, there is a menu with different options that we can use during the session:

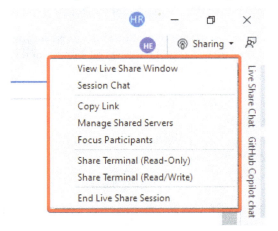

Figure 10.8 – The Live Share menu during a live session

Let's review these options and functionalities further:

- **View Live Share Window**: View the session status, including the session details and collaborators in the session.

- **Session Chat**: Open a chat window, where session participants can share comments.

- **Copy Link**: Copy the link in the clipboard to share it with others.

- **Manage Shared Servers**: Open a new window where we can share our local server with other users in the session.

- **Focus Participants**: Return participants to the code currently being focused on by the host.

- **Share Terminal (Read-Only)**: Share the terminal in read-only mode to share the console log and the results with others. Collaborators cannot execute any command in the terminal.

- **Share Terminal (Read/Write)**: Share the terminal with others in the session, with the possibility to execute commands remotely.

- **End Live Share Session**: End the session for all the connected users.

Let's click **View Live Share Window** to see the session status. It will show a new window in the right panel, as we can see in *Figure 10.9*:

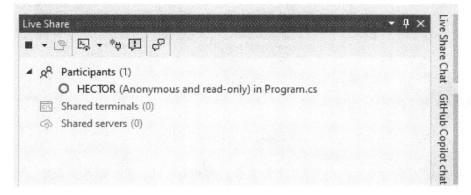

Figure 10.9 – A Live Share window during a session

In the **Participants** section, we can see the guest user who is watching the code in Program.cs. We can also see the shared terminals and servers. By clicking the red square in the upper-left corner, we can end the session at any time for all the participants.

> **Important note**
>
> Guest users join the session in read-only mode, which means that they cannot modify or update the code. This is a security requirement. To read more about security in Live Share, visit `https://docs.microsoft.com/en-us/visualstudio/liveshare/reference/security`.

It's quite simple to create and share a session using Live Share. We can see the collaborators connect and end the session at any time. Let's see how to edit or update code in real time.

Performing live editing

There are several ways to allow users joined to a Live Share session to edit source code. One of them is to enable write permission for a specific user by selecting the user and clicking the **Make read-write** option:

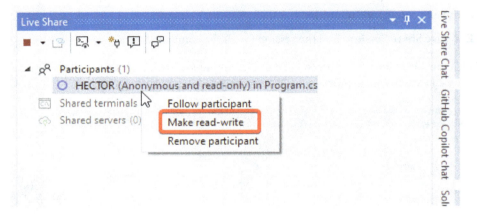

Figure 10.10 – Allowing a collaborator to edit

A second way is for the user to join VS 2022 through a previously logged-in account. If the user is in VS Code on the web, they can perform the exchange from **Menu | Open in Visual Studio**.

After selecting **Open in Visual Studio**, VS will load the project and show a loading message, eventually showing the code with the active session. We will see a menu where the Live Share button is located to manage the session as a collaborator (see *Figure 10.11*):

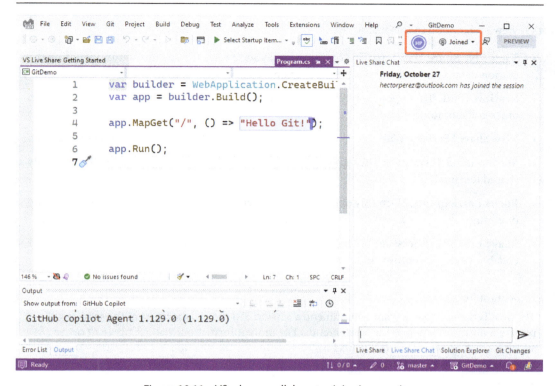

Figure 10.11 – VS when a collaborator joins in a session

We can click on **Joined** to see the options that we have as a collaborator in an active session (see *Figure 10.12*):

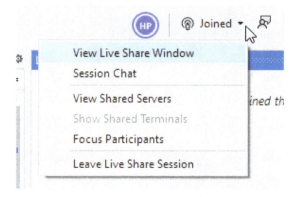

Figure 10.12 – The Live Share menu for collaborators

Let's review each option in this menu:

- **View Live Share Window**: This option allows you to view the session status, including the session details and collaborators in the session

- **Session Chat**: This opens a new window where you can type and share messages with other collaborators

- **View Shared Servers**: With this option, you can visualize the shared servers in the current session

- **Show Shared Terminals**: This option opens a new window where you can see the currently shared terminals

- **Focus Participants**: With this option, you will focus on what other collaborators are doing or editing

- **Leave Live Share Session**: With this option, you will leave the current session, while other collaborators can continue in the session

> **Important note**
>
> The host is the only participant who can end a session. All the collaborators in the session will receive a notification when the session ends. For more information, visit `https://docs.microsoft.com/visualstudio/liveshare/use/share-project-join-session-visual-studio`.

Now, you can edit any part of the code and see how Live Share works. For example, in *line 4* of the `Program.cs` file, we can add a comment for the `MapGet` method (see *Figure 10.13*):

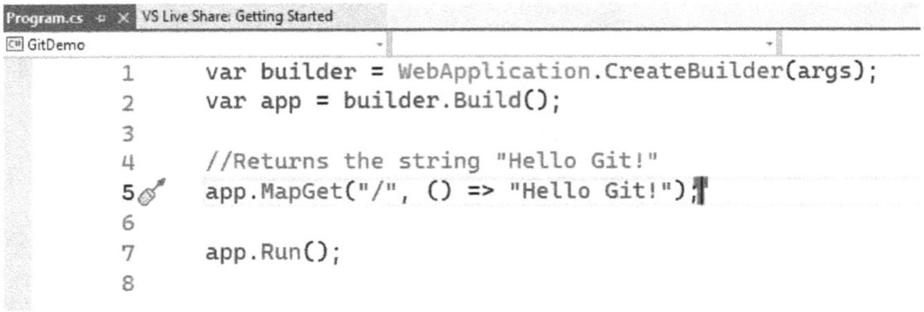

```
Program.cs  ⊞ X  VS Live Share: Getting Started
GitDemo
     1    var builder = WebApplication.CreateBuilder(args);
     2    var app = builder.Build();
     3
     4    //Returns the string "Hello Git!"
     5    app.MapGet("/", () => "Hello Git!");
     6
     7    app.Run();
     8
```

Figure 10.13 – Adding a comment during a Live Share session

Once the collaborator starts typing, the host and other users in the session can see the changes in the file in real time. The collaborator's name will be displayed in the specific line, highlighted with a random color (see *Figure 10.14*):

```
Program.cs ⇌ ✕
GitDemo
()        1    var builder = WebApplication.CreateBuilder(args);
          2    var app :WebApplication = builder.Build();
          3
          4    //Returns the string "Hello Git!"
          5    app.MapGet( pattern: "/",  handler: ()hector
          6
          7    app.Run();
          8
```

Figure 10.14 – The host watching changes in real time made by a collaborator

> **Important note**
> Live Share will assign a color for each new collaborator randomly to easily identify each user during the session.

You can edit the file to perform a suggestion, but you can also save the file using the **File | Save Selected Items** menu or the *Ctrl + S* shortcut.

Now that we know how Live Share works and some of the features that we can use during a shared session, we are ready to review an option to share a terminal with other collaborators during a session.

Sharing a terminal with other collaborators

In Live Share, we can also share a terminal with other developers. By sharing the terminal, we allow other developers to see more details of a project using the command line. After creating a new session, we can use the **Share Terminal** options to allow others to use our terminal (see *Figure 10.15*):

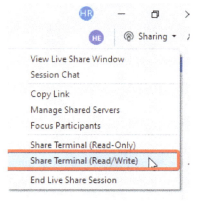

Figure 10.15 – The Share Terminal options in the Live Share menu

After clicking on **Share Terminal (Read/Write)**, we will see a new terminal in a window with the text [**Shared**] above the window title, indicating that the window is being shared. In *Figure 10.16*, we can see the new terminal added in VS in the bottom panel:

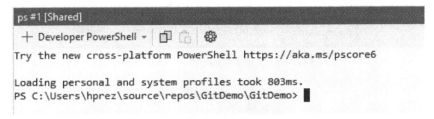

Figure 10.16 – A terminal shared during a Live Share session

Other collaborators will see this shared terminal automatically when they join the session. The collaborator in the session can execute any command in the project to get more details or add information using the Windows commands. In this case, we can use the .NET **command-line interface (CLI)**, which is a tool included in .NET, where we can compile, restore, run, and deploy .NET-based projects, using simple commands in the terminal.

To interact with the project, you can compile it using the dotnet build command. To learn more about the .NET CLI, you can check out the documentation at the following link: https://docs.microsoft.com/dotnet/core/tools/.

We will see warnings and errors during the compilation process in the terminal (see *Figure 10.17*):

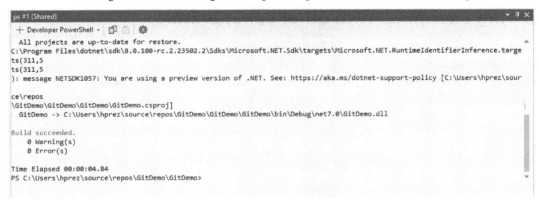

Figure 10.17 – A collaborator executes the dotnet build command in the shared terminal

The host and other collaborators in the session can see all the commands executed and the results in the terminal. The command is executed in the host's environment, that is, in the environment of the host that created the session.

If the terminal is not displayed by default or you want to see which other terminals are shared in the session, you can go to the **Live Share** window and see the status of the current session, including the shared terminals and the permissions for them (see *Figure 10.18*):

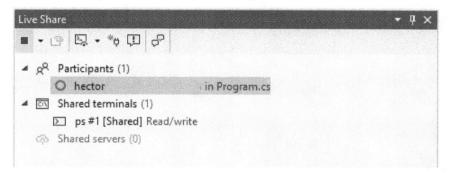

Figure 10.18 – The Live Share window with a shared terminal

Now that you know how to use Live Share in VS 2022 as a host and collaborator, you can invite others to collaborate on your project and share a terminal when it's required.

Summary

Live Share is an amazing tool for collaborating with others in real time. With globalization, this tool is now more important than ever, considering that a global development team can include developers working around the world in different time zones and with different tools.

You have learned how to use Live Share to work as a team and collaborate with other developers on the same project. You know how to load a Live Share session in VS, see the participants in the session, share a terminal, and end the session whenever you want. Also, you reviewed some alternatives to Live Share with other IDEs and editors.

In *Chapter 11*, *Working with Extensions in Visual Studio*, you will study how to increase the tools and functionalities included by default in VS using extensions. You will learn how to search, install, and set up extensions in VS and how to take advantage of them to increase your productivity. Once you have learned the process, you will create a simple extension.

11

Working with Extensions in Visual Studio

We cannot deny that VS's native functionality for performing tasks, as we have seen so far, is phenomenal. However, there may be times when you want to extend the capabilities of the IDE with simple features, such as applying a new theme with custom colors, or complex functionality, such as code refactoring tools.

It is in these cases where VS extensions are of great help, which is why we will dedicate an entire chapter to analyzing their use.

The main topics we will cover are as follows:

- Working with the **Extensions** tool
- Searching for and installing extensions
- Reviewing VS Marketplace
- Setting up extensions
- Creating a VS extension

Let's start exploring the extensibility of VS, which is due to extensions.

Technical requirements

As this chapter focuses on showing the use of extensions in VS, a specific code repository is not required; however, you can have the repository located at `https://github.com/PacktPublishing/Hands-On-Visual-Studio-2022-Second-Edition/tree/main/Chapter%207` at hand to make comparisons throughout this chapter.

Let's learn how to work with the **Extensions** tool from VS.

Working with the Extensions tool

The main purpose of extensions in VS is that you can improve your day-to-day productivity with features that may be somewhat specific to particular tasks, such as providing suggestions for best practices in code, performing code cleanup, highlighting messages in the output window, adding visual features in code files, or interacting with SQLite databases. Allowing developers to create new features for VS and share them with the rest of the world is an excellent move on the part of the Microsoft team.

> **Important note**
>
> In this section, I will show you simple extensions to familiarize you with the concept of extensions. In *Chapter 12, Using Popular Extensions*, you will learn about the most used and preferred extensions by developers and how to work with them.

But how can we access these extensions? The most direct answer is through the **Extensions** tool, which you can find through the **Extensions | Manage Extensions** menu, as shown in *Figure 11.1*:

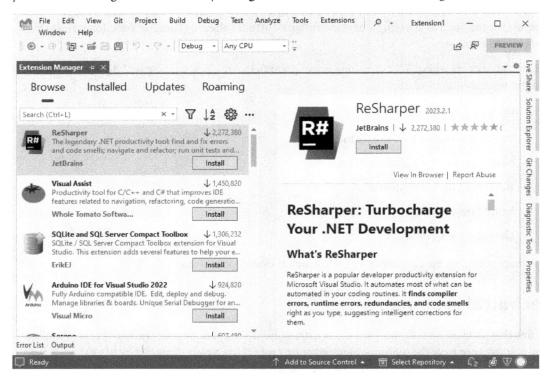

Figure 11.1 – The Extensions tool

In the **Extension Manager** window, we will see the extensions listed, alongside the extension's name, author, rating, version, number of downloads, whether it's free or paid, a description, and the **View in Browser** link, which opens a browser for the extension's page in VS Marketplace, which provides more details (if a free trial is available, the Q&A, reviews by other developers, tags, categories, which VS version is supported, and so on). This will be discussed in the *Reviewing VS Marketplace* section.

In addition, we'll see a set of tabs that allow us to filter the extensions:

- **Browse**: In this category, you will find the extensions that can be added to VS. In the top part, we have different buttons to filter and sort the results.

- **Installed**: These are the extensions that we have previously installed in the VS instance.

- **Updates**: This category shows the extensions that have pending updates, either with bug fixes or extension improvements.

- **Roaming**: This category allows you to view extensions that were previously installed in an instance of VS that are associated with a Microsoft account so that they can be installed in the current instance of VS. This means that if you are logged in on a different computer and you install an extension, you will be able to see it in this list, so you do not have to remember the name of the extension and can install it with ease.

Now, let's see how we can search for and install extensions thanks to this tool.

Searching for and installing extensions

The **Extensions** tool, which is shown in *Figure 11.2*, has a search box located at the top, where we can enter a search term to find extensions referring to some technology or tool. It is important to note that the search will be performed in the selected tab, as we discussed in the *Working with the Extensions tools* section.

If you want to perform a search among all the extensions, the best thing to do is to go to the **Browse** category and perform the search. In our example, we'll search for extensions related to the keyword javascript:

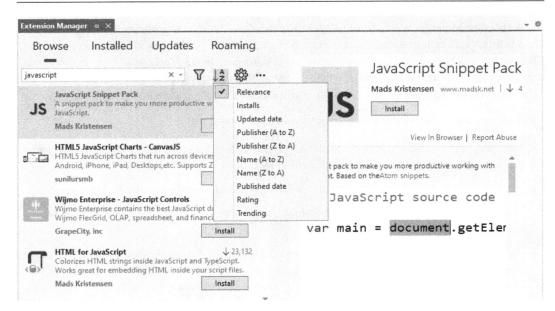

Figure 11.2 – A list of results in the Extensions tool

The order of the list of extensions will appear according to the drop-down item at the top, which appears as sorting by **Relevance** by default. However, we can also select other options, such as displaying by **Installs**, **Updated date**, **Trending**, or **Rating**.

Let's do another test, this time to install an extension. Let's look for extensions related to css. In the list, you will see an extension called **Color Preview** by **Mads Kristensen**. To install this extension, just click on the **Install** button; this will start the installation process:

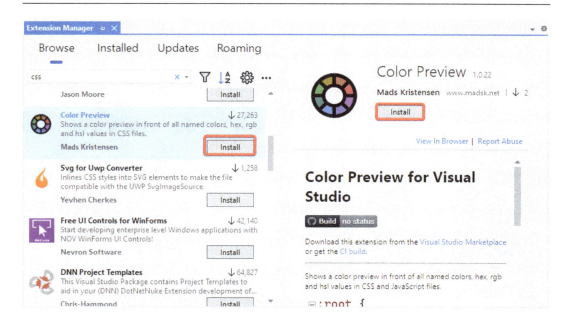

Figure 11.3 – Searching for and installing an extension

Once VS is ready to install the extension, the button text will change to **Cancel Install**. You must restart VS to perform the installation.

> **Note**
>
> Sometimes, you are likely to find extensions with similar names. To install the correct extension, it is advisable to check the name of the extension author, as well as other metadata.

Once we close VS, the installation process will start, which will show you (as shown in *Figure 11.4*) the extension that will be installed on the VS instance, as well as additional information, such as the **Digital Signature** and **License** types:

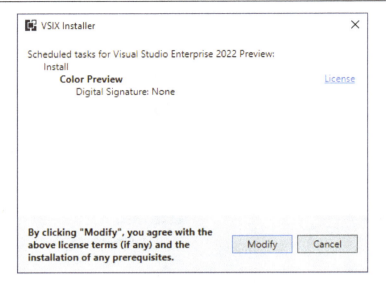

Figure 11.4 – The extension installation window

If we agree with the displayed information, as shown in *Figure 11.4*, we must click on the **Modify** button, which will start the installation process in the IDE.

Once the installation process is complete, we must reopen an instance of VS. Although it seems that nothing has changed at first glance, this extension will have added the ability to show us the selected color that was applied in the properties of a css file to the IDE, as demonstrated in *Figure 11.5*:

```css
site.css
 1  html {
 2      font-size: 14px;
 3      color: #FF1EF0;
 4      background-color: #1861ac;
 5  }
 6
 7  @media (min-width: 768px) {
 8      html {
 9          font-size: 16px;
10      }
11  }
12
13  .btn:focus, .btn:active:focus, .btn-link.nav-link:focus, .for
14      box-shadow: 0 0 0 0.1rem white 0 0 0 0.25rem #258cfb;
15  }
16
17  html {
18      position: relative;
```

Figure 11.5 – The feature added by the Color Preview extension

You can compare the preceding screenshot with those shown in the *Working with CSS styling tools* section of *Chapter 7, Styling and Cleanup Tools*.

If you want to uninstall an extension, just open the **Extensions** tool again and look for the extension in the **Installed** tab. Then, click on the **Uninstall** button, as shown in *Figure 11.6*:

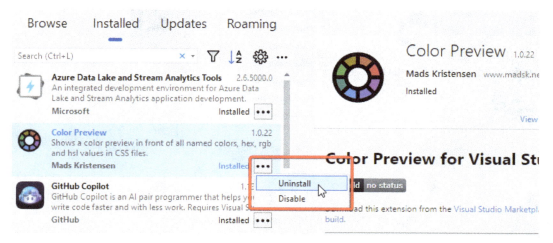

Figure 11.6 – Uninstalling an extension

In the preceding screenshot, you can see a second button called **Disable**, which allows you to temporarily disable an extension until you decide to reactivate it.

Now that we have seen how to install extensions via the **Extensions** tool, let's take a look at VS Marketplace.

Reviewing VS Marketplace

VS Marketplace is the online place to find and install extensions for VS 2022. In this marketplace, you can also find extensions for other products of the VS family, such as VS Code and Azure DevOps. You can access the marketplace via `https://marketplace.visualstudio.com/vs`.

Once you enter the portal, you will have a very different user interface to the **Extensions** tool, but the core operation is the same. In the main portal, you will be able to see the different extensions sorted by **Featured**, **Trending**, **Most Popular**, and **Highest Rated**.

To test the marketplace, let's search for the term `icons` and see the list of results. Although the list currently yields around 60 results, not all extensions are compatible with the most modern version of VS. Therefore, it is recommended to apply the filter for **Visual Studio 2022**, as shown in *Figure 11.7*:

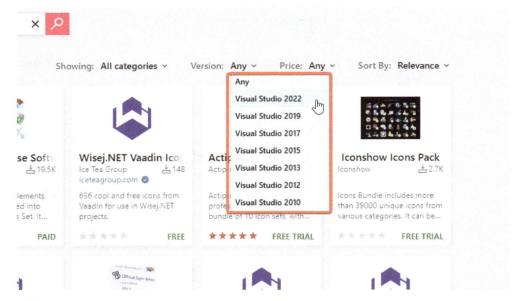

Figure 11.7 – Filtering VS versions in the marketplace

Once the filter has been applied, find the **Visual Studio Iconizer** extension and click on it to go to the extension's page:

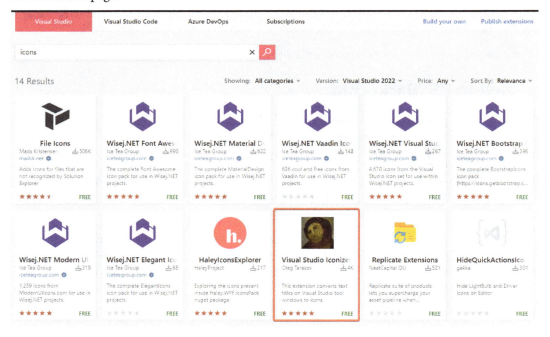

Figure 11.8 – Searching for extensions in the marketplace

On the extension's page, you will be able to find information that will give you a general idea about the extension, such as the number of installations, the number of reviews, ratings, the change log, and the project page, among other data. If the description of the extension specifies that it can help you solve a problem, you can click on the green **Download** button, as shown in the following screenshot:

Figure 11.9 – The button to download an extension

This will start the process of downloading a file with a `vsix` extension – in this particular case, the name is `VSIconizer.vsix`. To install the extension in VS, you must make sure you have closed all instances of the IDE and then run the downloaded file, which will start the same installation process that we saw in the *Searching for and installing extensions* section. Before installing the extension, the IDE looks like this:

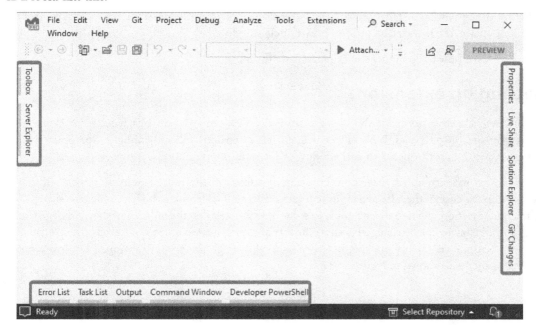

Figure 11.10 – The appearance of the IDE tabs before installing the Iconizer extension

Once the installation process has finished and an instance of VS 2022 has been opened, we can see how the appearance of the IDE tabs has changed, as illustrated in *Figure 11.11*, with the text being replaced by icons:

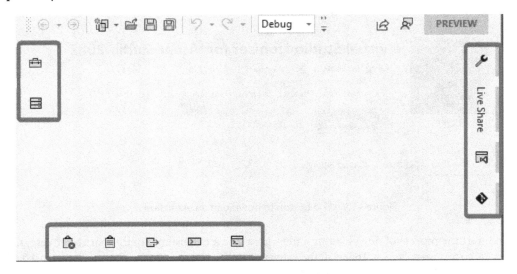

Figure 11.11 – The appearance of the IDE tabs modified by the Iconizer extension

Whether you opt for an installation with the **Extensions** tool or by downloading the extensions from the marketplace, the installation process is extremely simple.

Now, let's learn how to configure the extensions we install.

Setting up extensions

Unfortunately, there is no standardized way to configure a VS extension since each extension is unique and serves specific purposes. What is true is that most extensions will add configuration options for the extension, either through a special window or from the configuration options.

The best way to learn about these configuration options is through the extension page itself. For example, in the case of the **Visual Studio Iconizer** extension, which we installed in the *Reviewing VS Marketplace* section, the initial behavior is to show only the icons on the tabs. The extension page tells us that this behavior can be changed to show the text of the tab next to the icon added by the extension. This can be done via the options that have been added through the **Tools** | **Options** | **Environment** | **Iconizer** menu, as shown in *Figure 11.12*:

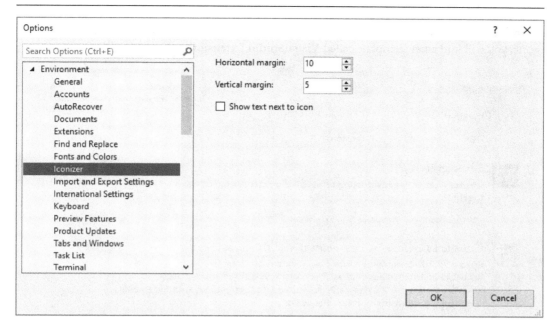

Figure 11.12 – The Iconizer extension options

Most extensions create a section, as shown in the preceding figure, and the more complex ones even provide a special new menu for the tool. We will discuss this in *Chapter 12, Using Popular Extensions*.

Now, let's learn how to create an extension for VS 2022.

Creating a simple VS 2022 extension

Now that you know how to install extensions in VS 2022, you may be wondering how to create an extension for VS. Although this process may sound very complex, fortunately, the team behind VS 2022 has been working on a new framework called **VisualStudio.Extensibility**. Although currently in the preview stage, it is the future as far as creating extensions for VS 2022 is concerned.

Some of the new features and characteristics of this framework are as follows:

- Improved performance and reliability

- A modern, asynchronous API

- Improved stability for testing

- A simplified architecture with consistent APIs

To create an extension using VisualStudio.Extensibility, you must search for and install the `VisualStudio.Extensibility Project System` extension via **Extension Manager**, as you saw in the *Searching for and installing extensions* section.

Once the new extension has been installed and VS has been restarted, in the template selector for new projects, you will find a new template called **VisualStudio.Extensibility Extension**:

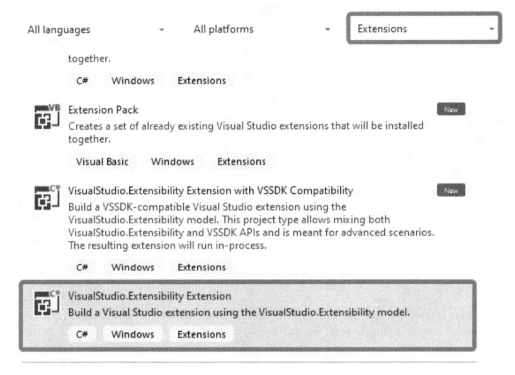

Figure 11.13 – The VisualStudio.Extensibility Extension template

Create a new project using this template using the name `ColorGeneratorExtension`. Once the project has been created, you will see that it consists of two files called `Command1.cs` and `ExtensionEntrypoint.cs`, as shown in *Figure 11.14*:

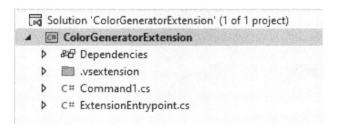

Figure 11.14 – Initial template files

Let's analyze the main components of each class. These are shown in *Figure 11.15*:

```
/// <summary>
/// Extension entrypoint for the VisualStudio.Extensibility extension.
/// </summary>
[VisualStudioContribution]
0 references
internal class ExtensionEntrypoint : Extension
{
    /// <inheritdoc/>
    0 references
    public override ExtensionConfiguration ExtensionConfiguration => new()
    {
        Metadata = new("ColorGeneratorExtension.6952a9f6-73d5-4213-b056-174e8ae42bf8",
            this.ExtensionAssemblyVersion,
            "Publisher name",
            "ColorGeneratorExtension"),
    };

    /// <inheritdoc />
    0 references
    protected override void InitializeServices(IServiceCollection serviceCollection)
    {
        base.InitializeServices(serviceCollection);

        // You can configure dependency injection here
        // by adding services to the serviceCollection.
    }
}
```

Figure 11.15 – The main components of the ExtensionEntryPoint class

The first one is `ExtensionEntrypoint.cs`. This class is the access point of the extension, which inherits from a base `Extension` class. The `Extension` class is abstract and contains methods that we can override to customize the extension. An example of this is the default overridden members – that is, the `ExtensionConfiguration` property, which defines the extension metadata ID, version, publisher name, and display name. A method called `InitializeServices` has also been overwritten, which allows us to configure dependency injection by adding services to be used during the extension life cycle. Finally, we can see the `VisualStudioContribution` attribute on the `ExtensionEntryPoint` class. This attribute allows us to indicate that this extension may be consumed by VS.

Now, let's review and modify the `Command1.cs` class, which is where we will be able to add the functionality of the extension. Let's start by renaming the class name, selecting the file in the Solution Explorer, and then pressing the *F2* shortcut. Change its name to `GenerateColorCommand.cs`, as shown in *Figure 11.16*:

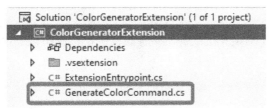

Figure 11.16 – The name of the modified command file

If you are asked whether you want to rename all code references, click on the **Yes** button. Now, if we open the `GenerateColorCommand` class, we will find the following code snippet:

```
[VisualStudioContribution]
internal class ColorGeneratorExtension : Command
{
    private readonly TraceSource logger;

    public ColorGeneratorExtension(VisualStudioExtensibility
extensibility, TraceSource traceSource)
        : base(extensibility)
    {
        this.logger = Requires.NotNull(traceSource,
nameof(traceSource));
    }
    public override CommandConfiguration CommandConfiguration =>
new("%ColorGeneratorExtension.Command1.DisplayName%")
    {
        Icon = new(ImageMoniker.KnownValues.Extension, IconSettings.
IconAndText),
        Placements = new[] { CommandPlacement.KnownPlacements.
ExtensionsMenu },
    };
    public override Task InitializeAsync(CancellationToken
cancellationToken)
    {
        return base.InitializeAsync(cancellationToken);
    }
    public override async Task ExecuteCommandAsync(IClientContext
context, CancellationToken cancellationToken)
    {
        await this.Extensibility.Shell().ShowPromptAsync("Hello from
an extension!", PromptOptions.OK, cancellationToken);
    }
}
```

Let's analyze the class members as we customize the extension. First, we can see that the class has the `VisualStudioContribution` attribute, which means that it can be used by VS.

Unlike the `ExtensionEntrypoint` class, the `ColorGenerationExtension` extension inherits from the `Command` class, which represents an action that can be initiated by the users, such as selecting a menu item, pressing a button on the toolbar, and so on.

We can see that a constructor has also been created that receives a parameter of the `VisualStudioExtensibility` type that allows communication with the IDE.

Another member that has been overwritten is the CommandConfiguration property, which is initialized by passing a displayName value. You will notice that a syntax with %% is used, which indicates that we will get the value from the string-resources.json file, which is located inside the .vsextension folder.

Let's get to work and change the value inside the string-resources.json file to the following:

```
{
"ColorGeneratorExtension.GenerateColorCommand.DisplayName": "Color
Generator"
}
```

When making this change, we must also modify the initialization of the CommandConfiguration property so that the name matches the value of the JSON file, as follows:

```
public override CommandConfiguration CommandConfiguration =>
new("%ColorGeneratorExtension.GenerateColorCommand.DisplayName%")
```

As you can see, as part of the initialization of this property, the Icon and Placements properties are also being initialized. The Icon property allows us to assign an icon to be displayed in the command, while the Placements property specifies the place where the command will appear – in this case, in the **Extensions** menu.

Let's change the behavior of the command by modifying some of its properties, as shown here:

```
public override CommandConfiguration CommandConfiguration =>
    new("%ColorGeneratorExtension.GenerateColorCommand.DisplayName%")
{
    Icon = new(ImageMoniker.KnownValues.ColorPalette,
        IconSettings.IconAndText),
    Placements = new[]
    {
        CommandPlacement.KnownPlacements.ExtensionsMenu
    },
    TooltipText =
        "Generate a new random hexadecimal value",
    Shortcuts = new CommandShortcutConfiguration[]
    {
        new(ModifierKey.LeftAlt, Key.M),
    },
    VisibleWhen =
        ActivationConstraint
        .ClientContext(ClientContextKey
                        .Shell
                        .ActiveEditorContentType, ".+"),
};
```

With the preceding code, I have modified the command icon, added a tooltip to let users know what the command does, and added a shortcut to execute the command quickly. I have also specified a rule to determine that the command will only be visible when the user is working in a text editor window.

Another method that has been overwritten by default is the `InitializeAsync` method, which allows a one-time configuration to be performed when initializing the command.

The last method that is overwritten is the `ExecuteCommandAsync` method, which receives `IClientContext` and `CancellationToken` values. This method is where we define what the command will do when it is executed. By default, it will show the text **Hello from an extension!**

Let's modify the command to make the extension more useful by allowing us to insert a color in a randomly generated hexadecimal format. We can do this with the following code snippet:

```
public override async Task ExecuteCommandAsync(
    IClientContext context,
    CancellationToken cancellationToken)
{
    Requires.NotNull(context, nameof(context));

    var random = new Random();
    var colorHexString =
        $"#{random.Next(0x1000000):X6}";

    using var textView =
        await context
            .GetActiveTextViewAsync(cancellationToken);
    if (textView is null)
    {
        logger
            .TraceInformation("There was no active text view when
command is executed.");
        return;
    }

    var document = textView.Document;
    await this.Extensibility
        .Editor()
        .EditAsync(
        batch =>
        {
            document
            .AsEditable(batch)
            .Replace(textView.Selection.Extent,
                colorHexString);
```

```
        },
        cancellationToken);
}
```

Here are the main steps of this code:

1. We ensure that the `context` object is not null using the `Requires.NotNull` method.

2. A hexadecimal color is generated and stored in the `colorHexString` variable.

3. An attempt is made to obtain the active text view through the `GetActiveTextViewAsync` method. If none is found, an information log is created and the execution of the command is terminated.

4. If a text view has been found, the next step is to obtain the `Document` associated with the text view.

5. We access the `Extensibility` property, to which we have access thanks to the fact that we inherit from the `Command` class, and which provides functionality to interact with the VS environment.

6. The `Editor()` method is executed, which returns the object representing the VS editor. The obtained object allows the editor to be interacted with so that edits can be made.

7. The `EditAsync()` method is used to initiate a batch edit on the document.

8. Within the Lambda expression inside `EditAsync`, the `AsEditable()` method is used, which enables the document to be edited and the hexadecimal color to be written.

Once the command has been declared, we must change the way the extension is displayed in the toolbar, as shown in *Figure 11.17*, by selecting the **Current Instance** [your VS edition] option – for example, **Current Instance [Visual Studio Enterprise 2022]** – so that we can debug the extension in case we need to do so:

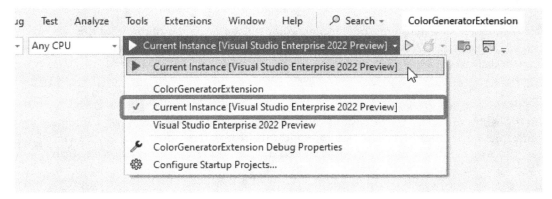

Figure 11.17 – Selecting the instance to start debugging

Once you have selected the **Current Instance [Visual Studio Enterprise 2022]** option, start debugging the extension. This will open a new VS instance called **VS Experimental Instance**, which is a clean VS instance for testing.

As you may recall, we indicated that the extension was to be found in the **Extensions** menu; however, when you open this option, you will see that the extension does not appear there:

Figure 11.18 – The Extensions menu without the extension created

This happens because we specified a rule through `.ActiveEditorContentType, ".+"`, to display the extension only when a text editor is available. To display the extension, go to **File | New | File**, select **Text File** (although you could select a C# file, HTML file, and so on), and click the **Open** button.

If you open the new file in the IDE and go to the **Extensions** section, you will see that the extension we have created is already displayed:

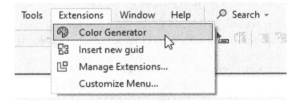

Figure 11.19 – The Extensions menu showing that the extension has been created

After clicking on the extension, a hexadecimal color will appear in the editor. With this, we have created a simple extension. At this point, you have the basis to create your own extension to help you with your daily tasks within VS.

Note

It is possible to publish the extensions you've created in VS Marketplace. If you want more information about this process, you can follow the instructions at `https://learn.microsoft.com/en-us/visualstudio/extensibility/walkthrough-publishing-a-visual-studio-extension?view=vs-2022`.

Summary

In this chapter, we have seen that extensions are a way in which we can extend the functionality of VS to improve our user experience and optimize development time. We have tested a few extensions that have completely changed some of the IDE's functionality, such as previewing colors in .css files, renaming tabs to icons, and creating themes for distribution.

We have also reviewed how to search for and install extensions, both from the **Extensions** tool and VS Marketplace. We analyzed how extensions are regularly configured, and finally, we created a new extension that has allowed you to experience the process that's involved in creating extensions with the new Visual.Studio Extensibility framework.

In *Chapter 12, Using Popular Extensions*, we will delve even deeper into the topic of extensions, analyzing which ones are the most popular because of their usefulness in daily development.

Using Popular Extensions

In *Chapter 11*, *Working with Extensions in Visual Studio*, we learned how to extend utilities and functionalities by installing extensions published by the VS community and third-party vendors. We can install these extensions using **Extension Manager** in VS and then restart VS to see the changes in the user interface.

In this chapter, we will analyze some free and useful extensions to increase productivity and improve our experience of using VS. We will install and review these extensions in the following sections:

- Using HTML Snippet Pack
- Cleaning up code with CodeMaid
- Compiling web libraries with Web Compiler
- Identifying white space with Indent Guides
- Optimizing images with Image Optimizer
- Highlighting messages in the **Output** window with VSColorOutput64
- Managing `.resx` files easily with ResXManager

We will start with HTML Snippet Pack, an extension that helps us include some additional code snippets in VS when coding HTML files.

Technical requirements

Since this chapter focuses on showing the functionality of various extensions, a project is not required for this chapter.

Using HTML Snippet Pack

In *Chapter 4*, *Adding Code Snippets*, we reviewed how code snippets can improve our productivity while we are coding. We also learned how to create, import, and remove code snippets using **Code Snippets Manager**.

In VS's extension marketplace, we can find many extensions to add code snippets for different technologies by navigating to `https://marketplace.visualstudio.com/` and typing `snippet` in the search bar (see *Figure 12.1*):

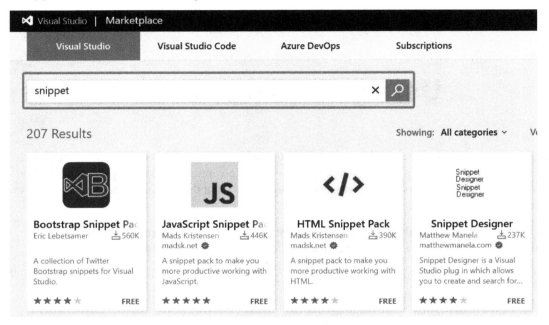

Figure 12.1 – Extensions related to snippets in VS Marketplace

One of the most popular snippet packs for web developers is **HTML Snippet Pack** by **Mads Kristensen**. With this extension, you can code in HTML faster, creating pieces of code and HTML elements after typing some characters in the editor.

To install the extension, search for the term `HTML Snippet` in the **Extensions** menu, as shown in *Chapter 11*, *Working with Extensions in Visual Studio*.

To use the HTML Snippet Pack extension, you need to navigate to an HTML file, which you can do quickly by going to **File | New | File** and selecting the **HTML Page** template. This will create a new HTML file in the editor.

Within this file, we can type the word `article` in the body element (see *Figure 12.5*):

```
<!DOCTYPE html>

<html lang="en" xmlns="http://www.w3.org/1999/xhtml">
<head>
    <meta charset="utf-8" />
    <title></title>
</head>
<body>
    article
</body>
</html>
```

Figure 12.2 – Typing article in index.html

After typing the word `article`, press *Tab* on your keyboard to easily generate the HTML element for an article in the related part of the code. *Figure 12.3* shows the article element generated automatically:

```
1     <!DOCTYPE html>
2
3     <html lang="en" xmlns="http://www.w3.org/1999/xhtml">
4     <head>
5         <meta charset="utf-8" />
6         <title></title>
7     </head>
8     <body>
9         <article></article>
10    </body>
11    </html>
```

Figure 12.3 – The article generated using a code snippet

Just as we generated the `article` element, we can easily create elements for `li`, `ul`, `img`, `input`, and almost all the existing elements in the HTML standard.

Read more

You can read more information about HTML Snippet Pack on the official website and repository at `https://github.com/madskristensen/HtmlSnippetPack`.

Now, let's review a different extension to analyze and format our code.

Cleaning up code with CodeMaid

CodeMaid VS2022, by **Steve Cadwallader**, is an amazing extension that helps us simplify and clean up code. It is free and compatible with C#, C++, F#, VB, PHP, PowerShell, R, JSON, XAML, XML, ASP, HTML, CSS, LESS, SCSS, JavaScript, and TypeScript.

To install CodeMaid, type codemaid in the search bar of the **Extensions** window, as shown in *Chapter 11, Working with Extensions in Visual Studio*, and install the **CodeMaid VS2022** version.

After installing CodeMaid, you will see a new option in the **Extensions** menu that contains all the functionalities and configurations related to CodeMaid (see *Figure 12.4*):

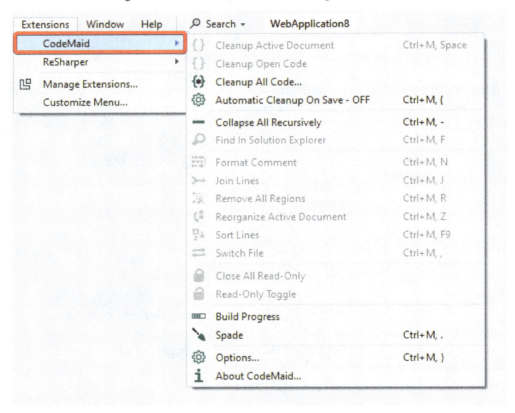

Figure 12.4 – The new CodeMaid option in the Extensions menu

For this demonstration, I'm going to use a new project with the **ASP.NET Core Web App (Razor Pages)** template. When there is no active document, many options are disabled, as shown in *Figure 12.4*; however, we can use the **Cleanup All Code…** option to perform a cleanup on the whole project, using the default settings in CodeMaid. We will get a confirmation message before starting the process. Click on **Yes** to continue (see *Figure 12.5*):

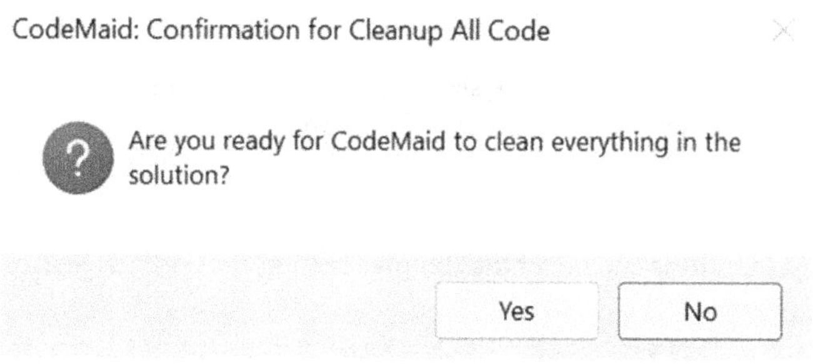

CodeMaid: Confirmation for Cleanup All Code   ✕

?   Are you ready for CodeMaid to clean everything in the
solution?

Yes   No

Figure 12.5 – Confirmation to clean up the project using CodeMaid

After confirming, CodeMaid will analyze each file at a time, look for white spaces and empty lines, and sort lines, among other things. You will see a progress bar and the current file being processed (see *Figure 12.6*):

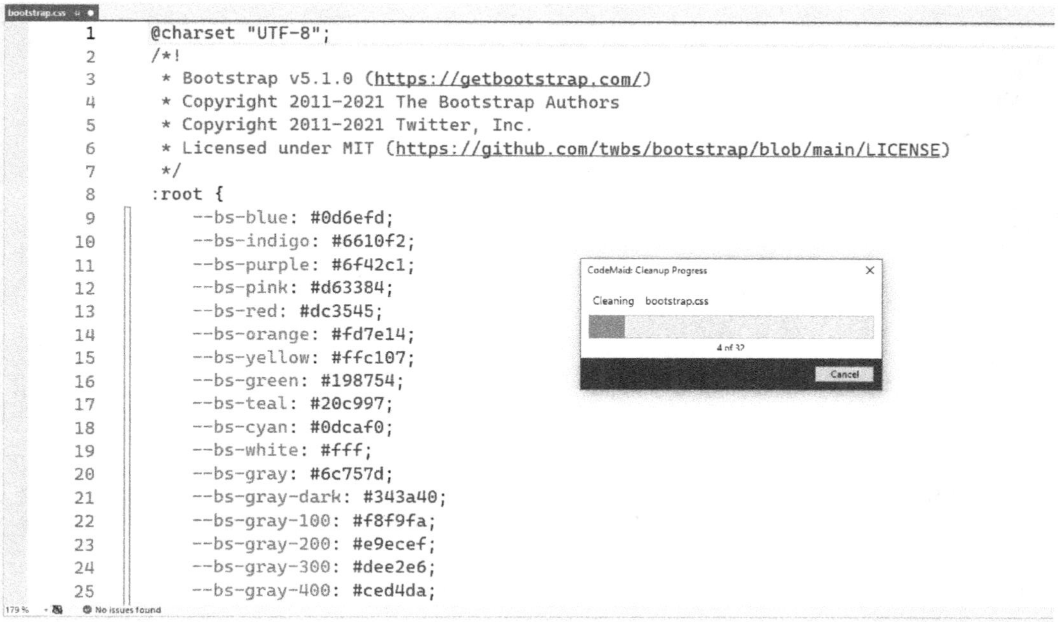

```css
bootstrap.css
1   @charset "UTF-8";
2   /*!
3    * Bootstrap v5.1.0 (https://getbootstrap.com/)
4    * Copyright 2011-2021 The Bootstrap Authors
5    * Copyright 2011-2021 Twitter, Inc.
6    * Licensed under MIT (https://github.com/twbs/bootstrap/blob/main/LICENSE)
7    */
8   :root {
9       --bs-blue: #0d6efd;
10      --bs-indigo: #6610f2;
11      --bs-purple: #6f42c1;
12      --bs-pink: #d63384;
13      --bs-red: #dc3545;
14      --bs-orange: #fd7e14;
15      --bs-yellow: #ffc107;
16      --bs-green: #198754;
17      --bs-teal: #20c997;
18      --bs-cyan: #0dcaf0;
19      --bs-white: #fff;
20      --bs-gray: #6c757d;
21      --bs-gray-dark: #343a40;
22      --bs-gray-100: #f8f9fa;
23      --bs-gray-200: #e9ecef;
24      --bs-gray-300: #dee2e6;
25      --bs-gray-400: #ced4da;
```

CodeMaid: Cleanup Progress   ✕

Cleaning bootstrap.css

4 nf 32

Cancel

Figure 12.6 – A cleanup in progress using CodeMaid

After completing the cleanup, which includes removing extra spaces and lines, all the files will be saved.

> **Important note**
>
> Reducing lines in code means reducing the size of a file. Blank lines and white spaces make the code difficult to read and increase project size when published.

CodeMaid has some options that we can turn on or turn off, depending on our needs. Navigate to **Extensions | CodeMaid | Options** and select the **Remove** section to choose scenarios where CodeMaid can remove code (see *Figure 12.7*):

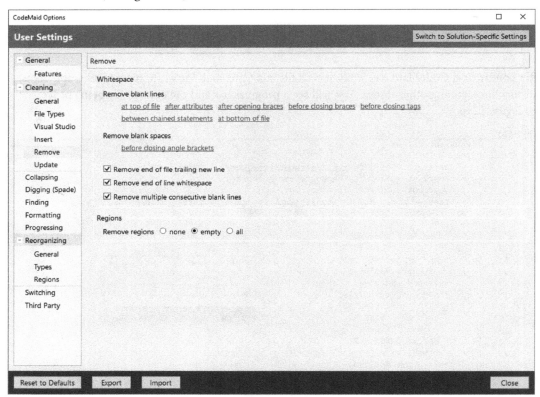

Figure 12.7 – User Settings for CodeMaid

CodeMaid also has many options related to removing blank lines and blank spaces. By default, all the options are enabled, but you can click on the options that you want to disable to set up CodeMaid according to the guidelines in your project.

> **Read more**
>
> You can read more information about CodeMaid on the official website: https://www.codemaid.net/.

Now that you know how CodeMaid works and can use it to clean up your projects, let's review the next extension to compile web files in VS.

Compiling web libraries with Web Compiler

If you are using TypeScript, LESS, or Sass in a project, you need to pre-compile the code to get the production version of your web project so that your browser can read every line of code. Note that your browser can only read CSS, HTML, and JavaScript. Using **Web Compiler** by **Jason Moore** in VS, you can do this easily and see the precompiled code directly.

Now that we know what Web Compiler is, let's install it and understand how to take advantage of this tool.

To install Web Compiler, go to **Extensions | Manage Extensions**, type web compiler in the search bar, and install the **Web Compiler 2022+** extension, as shown in *Chapter 11*, *Working with Extensions in Visual Studio*.

To use Web Compiler, we can navigate to and select any JavaScript file in any project and then right-click it. This will display the **Compile file** option in the menu, as shown in *Figure 12.8*:

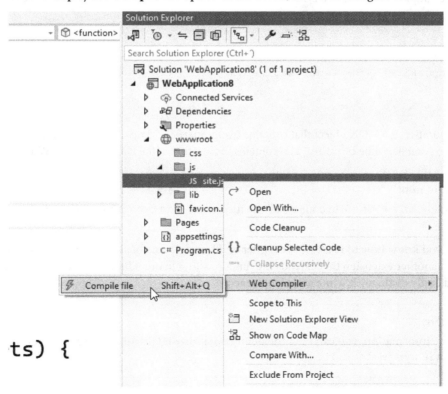

Figure 12.8 – The "Web Compiler" option in the project to compile a file

Using the **Web Compiler | Compile file** option, we can compile the file and generate a new version using **ECMAScript 2009 (ES5)**, which is a JavaScript specification that allows us to support old browser versions. See the file generated in *Figure 12.9*:

Figure 12.9 – site.es5.js generated by Web Compiler

Each file added to Web Compiler will be included in a file called compilerconfig.json. This file is associated with the compilation process with VS. This means that we can build and publish the project normally, and the files associated will be generated automatically.

Here is an example of a compilerconfig.json file after using it in the site.js file:

```
[
  {
    "outputFile": "wwwroot/js/site.es5.js",
    "inputFile": "wwwroot/js/site.js"
  }
]
```

Each configuration is a JSON object that contains two properties – inputFile is the location of the source file to compile, while outputFile contains the location of the file generated by Web Compiler.

> **Important note**
> You can use *Shift + Alt + Y* to compile all the files included in compilerconfig.json.

With that, you know how to use Web Compiler and how to transform JavaScript files to use ES5. Let's look at another extension that extends the functionalities in our editor and allows us to see some guides and easily distinguish white and blank spaces.

> **Read more**
> For more information, you can read the documentation on GitHub: https://github.com/failwyn/WebCompiler.

Identifying white spaces with Indent Guides

In *Chapter 11, Working with Extensions in Visual Studio*, we reviewed some extensions in VS, including **Color Preview**. With these extensions, we learned how the text editor in VS can be extended to improve our experience and provide more tools for some technologies and scenarios. **Indent Guides**, created by the user **Steve Dower [MSFT]**, is another example of this type of extension that extends the text editor in VS.

Indent Guides is a simple but useful extension that helps us identify extra white spaces and indentations in the structure of code.

To install Indent Guides, go to **Extensions | Manage Extensions** and type `Indent Guides` in the search bar. Then, install the **Indent Guides for VS 2022** extension, following the steps from *Chapter 11, Working with Extensions in Visual Studio*.

Once you have VS running again and the installation is completed, you can open any file of any programming language supported by VS 2022 and see new guides that show white spaces and tabs between text and the elements in the text editor (see *Figure 12.10*):

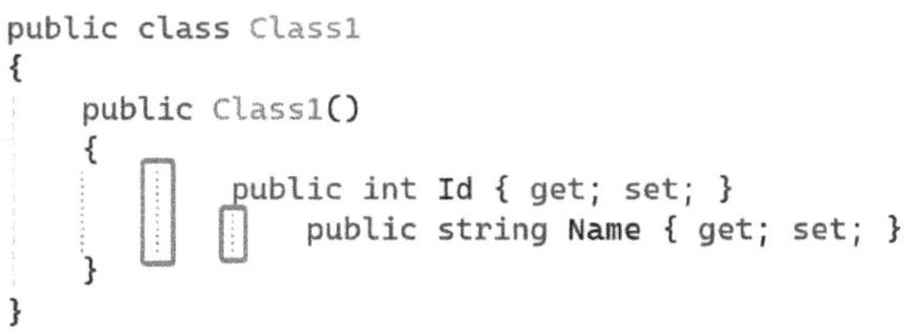

```
public class Class1
{
    public Class1()
    {
            public int Id { get; set; }
            public string Name { get; set; }
    }
}
```

Figure 12.10 – Guides in the VS text editor

This tool is amazing for improving the formatting in our code files. There are some additional options that we can adjust to fit our preferences. Navigate to **Tools | Options | Indent Guides**. There, we will find many options to change the appearance, behavior, and highlights and set a quick start or default configuration.

Now, I invite you to learn about an extension that will allow you to optimize the images of your projects.

Optimizing images with Image Optimizer

Images are an essential element in any software project. It is common that when we work with images, they have a large weight, either because we have downloaded stock photos, or because of the high resolution of today's mobile devices, and so on. This weight translates into a slower loading of pages in software projects, so it is essential to optimize them as best as possible.

Fortunately, the **Image Optimizer (64-bit)** extension, by **Mads Kristensen**, can help you optimize any JPEG, PNG, and GIF images. To install this extension, you must search for it in the **Extensions** tool with the term `Image Optimizer` and install it, as shown in *Chapter 11, Working with Extensions in Visual Studio*.

Once the extension has been installed, you can right-click on any image. You will see the following options:

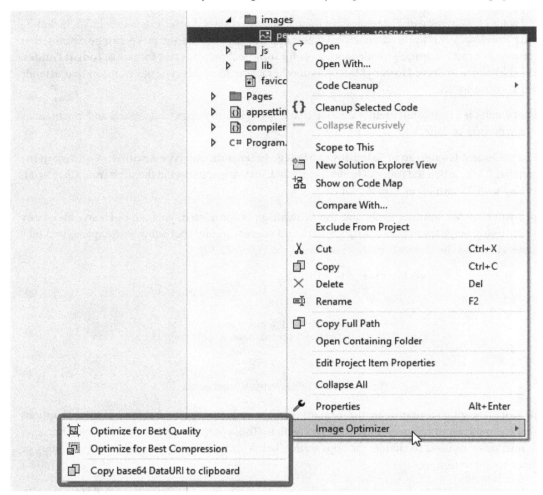

Figure 12.11 – Optimization options for images

In *Figure 12.11*, you can see the following options:

- **Optimize for Best Quality**: Optimizations will be made to the image without changing its quality

- **Optimize for Best Compression**: Optimizations will be performed by drastically reducing the image size but sacrificing a small part of the original quality (from the tests I have performed, this change is almost imperceptible to the human eye)

- **Copy base64 DataURI to clipboard**: The image will be copied to a base64 string

As an example, when performing the optimization run for the best compression of an image that originally weighs 2.99 MB, a reduction of 57.2% was obtained, as shown in the following output:

```
1 image optimized in 1.1 seconds. Total saving of 1797278 bytes / 57.2%

  purple-coneflower-8225677.jpg   optimized by 1797278 bytes / 57.2%
```

Figure 12.12 – Output window with optimization information

Undoubtedly, this plugin will help you to greatly optimize your projects of any kind without much effort.

Next, let's take a look at an extension that will be very useful when performing debugging.

Highlighting messages in the Output window with VSColorOutput64

If you have ever created a project with VS, you have probably had to use the **Output** window to print some result, or simply to see if there has been any error during the compilation or execution of your application.

Searching through the many lines generated in the **Output** window can be challenging since all the lines are the same color. The **VSColorOutput64** extension, by **Mike Ward – Ann Arbor**, can help us solve this problem. To install it, you must search for it with the term VSColorOutput64 in the **Extensions** window and install it, as described in *Chapter 11, Working with Extensions in Visual Studio*.

You can use any type of project to test this extension; in my case, I have created a console application with the following code:

```
using System.Diagnostics;
Debug.WriteLine("-- This is a test --");
Debug.WriteLine("Message: Hello World!");
Debug.WriteLine("Message: Héctor Pérez");
Debug.WriteLine("-- End of test --");
Console.ReadLine();
```

These lines are displayed in the **Output** window, alongside the following information:

```
Output
Show output from: Debug                              ▼ |    |      |   | ⚄ | ⚙ | ⏱
  ConsoleApp4.exe  (CoreCLR: clrhost): Loaded 'C:\Program Files\Microsoft Visual Studio\2022\Preview\Commo...
'ConsoleApp4.exe' (CoreCLR: clrhost): Loaded 'C:\Program Files\dotnet\shared\Microsoft.NETCore.App\8.0.0\
'ConsoleApp4.exe' (CoreCLR: clrhost): Loaded 'C:\Program Files\dotnet\shared\Microsoft.NETCore.App\8.0.0\
'ConsoleApp4.exe' (CoreCLR: clrhost): Loaded 'C:\Program Files\dotnet\shared\Microsoft.NETCore.App\8.0.0\
'ConsoleApp4.exe' (CoreCLR: clrhost): Loaded 'C:\Program Files\dotnet\shared\Microsoft.NETCore.App\8.0.0\
'ConsoleApp4.exe' (CoreCLR: clrhost): Loaded 'C:\Program Files\dotnet\shared\Microsoft.NETCore.App\8.0.0\
'ConsoleApp4.exe' (CoreCLR: clrhost): Loaded 'C:\Program Files\dotnet\shared\Microsoft.NETCore.App\8.0.0\
'ConsoleApp4.exe' (CoreCLR: clrhost): Loaded 'C:\Program Files\dotnet\shared\Microsoft.NETCore.App\8.0.0\
'ConsoleApp4.exe' (CoreCLR: clrhost): Loaded 'C:\Program Files\dotnet\shared\Microsoft.NETCore.App\8.0.0\
'ConsoleApp4.exe' (CoreCLR: clrhost): Loaded 'C:\Program Files\dotnet\shared\Microsoft.NETCore.App\8.0.0\
'ConsoleApp4.exe' (CoreCLR: clrhost): Loaded 'C:\Program Files\dotnet\shared\Microsoft.NETCore.App\8.0.0\
-- This is a test --
Message: Hello World!
Message: Héctor Pérez
-- End of test --
'ConsoleApp4.exe' (CoreCLR: clrhost): Loaded 'C:\Program Files\dotnet\shared\Microsoft.NETCore.App\8.0.0\
'ConsoleApp4.exe' (CoreCLR: clrhost): Loaded 'C:\Program Files\dotnet\shared\Microsoft.NETCore.App\8.0.0\
```

Figure 12.13 – Default Output window

At first glance, the information is lost between the debug lines. Now, to configure the VSColorOutput64 extension, you must go to **Tools | Options | VSColorOutput64**, from where you will be able to configure different options of the extension. In our example, we are interested in easily identifying patterns of the messages we have printed in the **Output** window. To do this, click on the configuration button of the **RegEx Patterns** option:

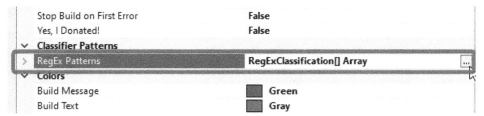

Figure 12.14 – Options to define RegEx Patterns

This newly opened window is named **RegEx Classification Collection Editor** and will show us a series of default regular expressions with which we can identify patterns in the **Output** window. We can add new records so that the extension can identify new patterns. For example, we have the following pair of messages:

- `-- This is a test --`

- `-- End of test --`

Both follow the same pattern, which begins with a pair of dashes, `--`, and ends with another pair of dashes, `--`, with text in between. A valid regular expression to identify this text would be `--\s*.*?\s*--`.

With this information, we will click on the **Add** option, which will add a new record. Here, we will select the **LogCustom3** option (although you can select the type of classification of your choice), set the **IgnoreCase** option to **True** to ignore upper- and lowercase letters, and enter the regular expression that I indicated before, as shown in *Figure 12.15*:

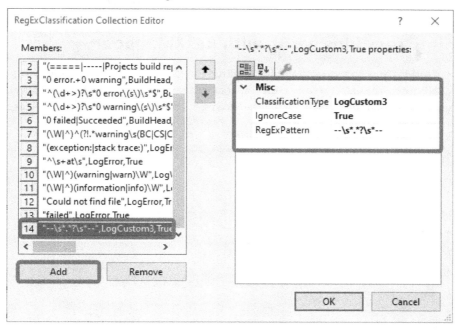

Figure 12.15 – Options to define RegEx patterns

After entering the necessary information and pressing the **OK** button on the floating windows, you will see how the messages in the **Output** window have changed color:

```
'ConsoleApp3.exe' (CoreCLR: clrhost): Loaded
'ConsoleApp3.exe' (CoreCLR: clrhost): Loaded
'ConsoleApp3.exe' (CoreCLR: clrhost): Loaded
'ConsoleApp3.exe' (CoreCLR: clrhost): Loaded
-- This is a test --
Message: Hello World!
Message: Héctor Pérez
-- End of test --
'ConsoleApp3.exe' (CoreCLR: clrhost): Loaded
'ConsoleApp3.exe' (CoreCLR: clrhost): Loaded
The program '[20632] ConsoleApp3.exe: Program
The program '[20632] ConsoleApp3.exe' has exit
```

Figure 12.16 – The Output window with custom pattern identification

In *Figure 12.16*, I have also created a regular expression to identify messages that start with the term `Message:` followed by text, which helps to identify them more easily. Undoubtedly, this extension can be of great help in identifying messages in the **Output** window.

Now, let's look at an extension that will allow you to manage the `.resx` resources of your projects easily.

Managing .resx files easily with ResXManager

In many applications, it is necessary to display localized information to users, to avoid errors in the use of the application. The use of **resource files**, also called `.resx` files, is one of the most common ways to display translations to users in multiple languages. VS has a basic `.resx` file editor, as shown in the following figure:

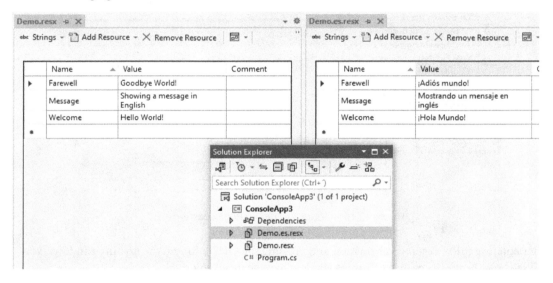

Figure 12.17 – The default VS .resx file editor

In *Figure 12.17*, you can see that the solution has a file named `Demo.resx` containing English strings and another file named `Demo.es.resx` for storing Spanish strings. In the same figure, you can see the basic `.resx` file editor, where each tab represents an open `.resx` file.

> **Note**
> If you want to learn more about resource files, you can visit the following link: `https://learn.microsoft.com/en-us/dotnet/core/extensions/create-resource-files`.

Imagine that you want your application to be displayed in more than two languages. This would imply creating multiple `.resx` files and opening them one by one each time you want to add a new value. This could result in errors as you would have to manually make sure you have the same number of lines in each `.resx` file and then translate each value. Fortunately, there is a free extension that will allow us to put an end to these problems.

Let's go to the extensions explorer, as shown in *Chapter 11, Working with Extensions in Visual Studio,* and then search for and install the **ResXManager** extension by **TomEnglert**.

Once you have installed the extension, you can open it from the **View | Other Windows | ResX Manager** menu. The **ResX Manager** option will open a new window, which will load all the `.resx` files recognized in the solution, showing valuable information such as the **Key** value and the different translations all in one place:

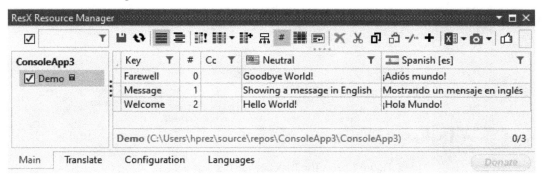

Figure 12.18 – The ResX Resource Manager extension window

To add a new string to the `.resx` files, simply click on the button with the + symbol. This will request the name of a new key:

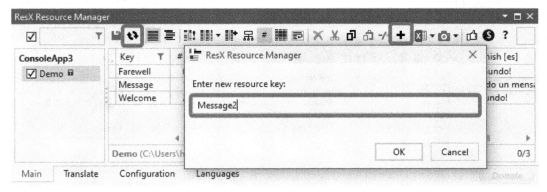

Figure 12.19 – Adding a new key to the .resx files

Finally, you can fill in the values for each language.

Undoubtedly, having all the records available in one place is a great help, but what if I told you that the extension can even help us to do the translations automatically? To experience this feature, you can go to the **Translate** tab in the bottom menu:

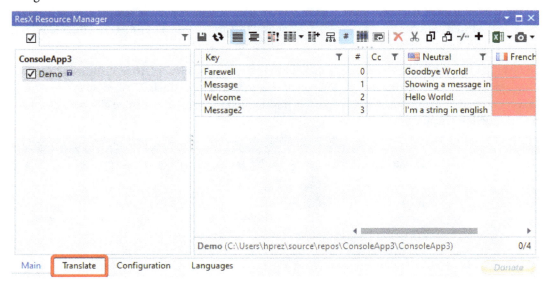

Figure 12.20 – The Translate tab highlighted

In the **Translate** window, you will see a list of several popular options that you can use for translations, such as **Azure Open AI**, **DeepL**, and others. In our case, we will use the **Google Lite** service, which does not require any previous configuration or keys to use it.

To run the test, I have added a new clean `.resx` file called `Demo.fr.resx` to handle the French strings, so we will get a window like the one shown in *Figure 12.21*:

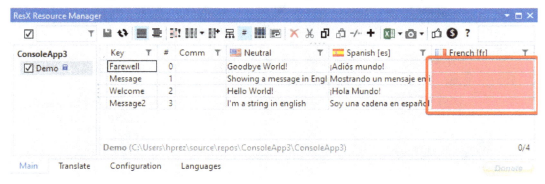

Figure 12.21 – Empty strings from the French string file

Important note

When adding or deleting a .resx file, you must press the refresh button shown in *Figure 12.19* so that the information in the table is updated.

Then, when you go to the **Translate** tab, you must select the **Google Lite** service, choosing the source language of the strings, as well as the target language, which in this case is French. You can finish this process by pressing the **Start** button, which will start the translation process for the strings:

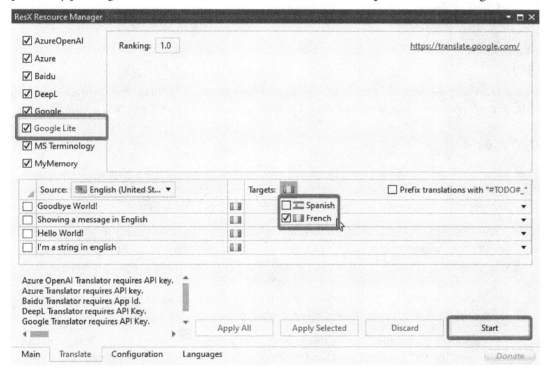

Figure 12.22 – String translation process

Once the translation process is finished, we must press the **Apply All** button to insert the changes in the selected .resx file:

Key ▼	#	Comm ▼	🌐 Neutral ▼	⬛ Spanish [es] ▼	🇫🇷 French [fr] ▼
Farewell	0		Goodbye World!	¡Adiós mundo!	Au revoir le monde!
Message	1		Showing a message in Engl	Mostrando un mensaje en i	Afficher un message en ang
Welcome	2		Hello World!	¡Hola Mundo!	Bonjour le monde!
Message2	3		I'm a string in english	Soy una cadena en español	je suis une chaîne en anglai

Figure 12.23 – Translated French strings

Finally, if you want to test the localization in a console application, you can do so by using the following code and changing the language to the one you want to test:

```
using Demo;
Thread.CurrentThread.CurrentUICulture = new CultureInfo("fr");
Console.WriteLine($"{Demo.Welcome}");
Console.WriteLine($"{Demo.Message}");
Console.WriteLine($"{Demo.Message2}");
Console.WriteLine($"{Demo.Farewell}");
Console.ReadLine();
```

ResXManager turns out to be an impressive extension with the great work of the developer community behind it.

This concludes our tour of some of the most popular and free extensions in **Visual Studio Marketplace**. I invite you to continue exploring the catalog of extensions and try the ones you find useful.

Summary

VS has a marketplace with many extensions that we can access using the VS **Manage Extensions** option. We can find many extensions related to code snippets in **Visual Studio Marketplace** and use HTML Snippet Pack to improve our productivity while coding in HTML files.

You now know how to use CodeMaid to clean up code and improve code quality in your projects. You can easily remove empty lines and white spaces and create a custom validation format for all the files in your project. You also learned how to install Web Compiler to compile and transform web files such as JavaScript files or libraries such as LESS and Sass into generic code that a browser can easily read. Then, you explored the Indent Guides extension, which shows us white and blank spaces in code to identify how to improve the format and structure of files.

You also learned how to optimize images using Image Optimizer, how to highlight messages in the **Output** window using VSColorOutput64, and how to easily manage string values located in .resx files using ResXManager.

In *Chapter 13*, *Learning Keyboard Shortcuts*, we will review the most important shortcuts included in VS by default. With this knowledge, you will be able to use a combination of some keyboard keys to perform common actions in VS.

13

Learning Keyboard Shortcuts

Throughout this book, we have analyzed some shortcuts that help us improve our productivity, using a combination of some keys to perform common actions in the IDE and source code. VS includes some useful shortcuts by default, but we can also create our own shortcuts, depending on our needs or common operations that we need to perform daily.

In this chapter, we will provide a summary of the most important shortcuts included by default in VS 2022 and explain how to create our own.

We will review the following topics in this chapter:

- The most-used shortcuts for use in source code
- The most common shortcuts for use in the IDE
- Creating custom shortcuts

When working with VS, you must know the relevant keyboard shortcuts so that you can carry out operations quickly. This will prevent you from wasting time on repetitive tasks, such as formatting a complete source code file or renaming a member of a class.

You must put these keyboard shortcuts into practice and begin to use them little by little, even if you execute them slowly at the beginning. You will see that with time, you will execute them automatically, without resorting to any visual aid.

Let's learn how to improve our productivity using shortcuts.

Technical requirements

Since this chapter focuses on showing shortcuts for use in any project, a base project is not necessary.

It is important to note that different keyboard mapping schemes can completely modify the shortcut keys. In addition, extensions such as **ReSharper** can modify keyboard shortcuts. To execute all the shortcuts shown in this chapter correctly, you must have a **Default** keyboard mapping scheme, which can be selected from the **Tools | Options | Environment | Keyboard** menu.

Let's enter the world of shortcuts, which will undoubtedly make you become a more efficient and productive programmer.

Shortcuts for use in source code

Working with source code involves working with thousands of lines of code, which can be a headache for even the most experienced developer. That is why the powerful search tools included in VS are an excellent way to search and navigate the lines of code.

Similarly, code editing and refactoring tasks are ongoing in projects, not to mention debugging and testing in large projects.

This is why working with shortcuts that give you instant access to these tools should be a priority in your career as a developer. Let's see what these keyboard shortcuts are.

Shortcuts for searching and navigating source code

Finding members of a class quickly may seem like a simple task when working on a one-class project, but you may not think so if you work with projects that contain hundreds of classes or even several projects within the same solution. It is during these moments that the following keyboard shortcuts become a great help.

Let's take a look at the shortcuts that can help us perform quick search and navigation operations in VS:

Shortcut	Description
Ctrl + Q	Opens **VS Search**
Ctrl + T	Opens the **Go To All** tool
Ctrl + 1 + T	Opens the **Go To All** tool with the **Type** filter
Ctrl + 1 + F	Opens the **Go To All** tool with the **File** filter
Ctrl + 1 + M	Opens the **Go To All** tool with the **Member** filter
Ctrl + 1 + S	Opens the **Go To All** tool with the **Symbol** filter
Ctrl + -	Goes back in opened documents of the current session
Ctrl + Shift + -	Goes forward in opened documents of the current session
F12	Goes to the class definition
Alt + F12	Opens the code of a class in a pop-up window for editing
Ctrl + F12	Goes to member implementation in a class
Ctrl + Shift + F12	Jumps to the next error in the error list when multiple errors exist
F8	Moves forward in the result list of the current window
Shift + F8	Moves backward in the result list of the current window

Table 13.1 – Shortcuts for quick search and navigation operations

The shortcuts discussed here allow you to quickly navigate between files, members, results, and implementations without having to take your hands away from the keyboard. Now, let's look at the most common shortcuts for editing and refactoring.

> **Important note**
>
> You can check out all the most common shortcuts in VS at `https://docs.microsoft.com/visualstudio/ide/default-keyboard-shortcuts-in-visual-studio?view=vs-2022`. We encourage you to download the file, print it, and keep it near you for quick reference.

Shortcuts for editing and refactoring

The commands that we will cover in this subsection correspond to those that allow you to apply changes directly to source code. Among the most common operations are renaming members, commenting on lines of code, and moving lines up and down.

Let's look at the shortcuts that make writing code even easier:

Shortcut	Description
Alt + Enter	Shows quick actions
Ctrl + .	Presents refactoring options
Ctrl + K, Ctrl + I	Retrieves class member information
Ctrl + K, Ctrl + C	Comments out several chosen lines in the code
Ctrl + K, Ctrl + U	Uncomments several chosen lines in the code
Ctrl + Shift + L	Eliminates selected lines
Ctrl + Shift + V	Shows clipboard history for pasting
Alt + ↑ / Alt + ↓	Moves code segment upwards/downwards
Ctrl + F	Searches for specific text in the code
Ctrl + A	Highlights every line in the current file
Ctrl + S	Commits the current file's unsaved modifications
Ctrl + Shift + S	Commits unsaved modifications in all open files
Ctrl + Shift + .	Enlarges the current file view
Ctrl + Shift + ,	Reduces the current file view
Ctrl + Up	Shifts selected lines upwards
Ctrl + Down	Shifts selected lines downwards
Ctrl + K, Ctrl + D	Applies formatting rules to the whole document
Ctrl + K, Ctrl + E	Executes code cleanup

Shortcut	Description
Ctrl + K, Ctrl + F	Applies formatting rules to selected lines
Ctrl + K, Ctrl + S	Wraps code with common structures
Ctrl + R, Ctrl + R	Changes a class member's name
Ctrl + R, Ctrl + E	Generates a property from a class field
Ctrl + R, Ctrl + G	Cleans unused imports and sorts them
Ctrl + R, Ctrl + M	Converts selected code into a method

Table 13.2 – Shortcuts for code writing

Now that we've learned about the main shortcuts for editing and refactoring code, let's take a look at those that help us optimize depuration and testing tasks.

Shortcuts for debugging and testing

Debugging and code execution are some of the most constant tasks we will perform while working with VS. Therefore, it is important to know the keyboard shortcuts that can help us to execute these tasks quickly. That is why, in this subsection, we will mention the most important shortcuts that focus on these tasks:

Shortcut	Description
F5	Initiates the application in debug mode
Ctrl + F5	Initiates the application without debug mode
Shift + 5	Halts the application during operation
Ctrl + Shift + F5	Halts the application, rebuilds the project, and initiates a new debug session
F9	Sets or removes a breakpoint
F10	Advances over code execution while debugging
F11	Debugs the source code step by step
Shift + F11	Exits the current method's execution
Ctrl + R, Ctrl + A	Begins unit test execution in debug mode
Ctrl + R, A	Begins unit test execution without debug mode

Table 13.3 – Shortcuts for coding tasks

This concludes the list of shortcuts that can help us to improve our time when working with source code.

Now, let's review the shortcuts that can help us perform quick actions in the IDE.

The most common shortcuts for use in the IDE

Knowing how to get around in the VS IDE through keyboard shortcuts is an important part of avoiding wasting time searching through menus to activate a specific panel. It is very common, for example, to close the **Solution Explorer** or **Properties** window by mistake and not know which menu contains the option to open them again. That is why, in this section, we will examine the shortcuts that speed up the performance of these tasks:

Shortcut	Description
Ctrl + [+ S	Selects an open file in the **Solution Explorer** window quickly
Ctrl + Alt + L	Activates the **Solution Explorer** window
Ctrl + Alt + O	Activates the **Output** window
Ctrl + \, E	Activates the **Error List** window
Ctrl + \, Ctrl + M	Activates the **Team Explorer** window
Ctrl + Alt + B	Activates the **Breakpoints** window
F4	Activates the **Properties** window
Alt + F6	Cycles backward through open-panel windows
Shift + Alt + F6	Cycles forward through open-panel windows
Shift + Esc	Closes the active tool window
Ctrl + Alt + Page Up	Scrolls up through open documents, irrespective of the session
Ctrl + Alt + Page Down	Scrolls down through open documents, irrespective of the session
Ctrl + Tab	Shows a window with open documents, highlighting the most recent
Ctrl + Shift + Tab	Shows a window with open documents, highlighting the least recent
Shift + Alt + Enter	Toggles full-screen mode in the VS environment for enhanced focus
Ctrl + K + K	Toggles a bookmark at the current line
Ctrl + K + N	Permits forward navigation among project bookmarks
Ctrl + K + P	Permits backward navigation among project bookmarks
Alt + G, C	Shows the window for Git modifications
Alt + G, M	Shows the window for Git repositories
Ctrl + Alt + F3	Reveals the selector for branches
Ctrl + Alt + F4	Activates the selector for repositories

Table 13.4 – Most common shortcuts for use in the IDE

> **Important note**
>
> Bookmarks are a feature of VS that allow you to mark lines in your code so that you can quickly return to them. You can find more information about them at `https://docs.microsoft.com/en-us/visualstudio/ide/setting-bookmarks-in-code?view=vs-2022`.

At this point, you have learned about the most common and useful shortcuts in VS. However, there is a way to create your own shortcuts so that you can adapt VS to your needs. Let's analyze how to create custom shortcuts.

Creating custom shortcuts

We can create shortcuts for specific actions in VS, and there are several options available to customize the current shortcuts.

You can navigate to **Tools | Options | Environment | Keyboard** to see all the current shortcuts in VS (as shown in *Figure 13.1*):

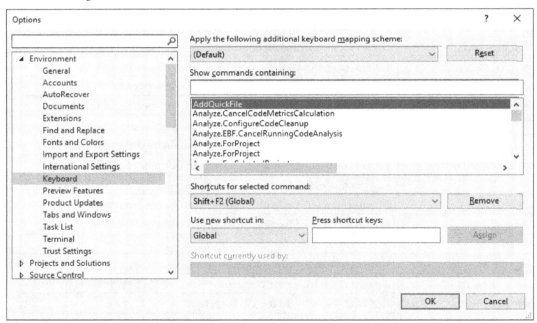

Figure 13.1 – The "Keyboard" option for customizing shortcuts

Here, you will find all the current shortcuts in VS for all the functionalities and a scheme for the shortcuts where you can set them up, depending on the context. By default, VS includes different keyboard schemes with different keyboard shortcut configurations (see *Figure 13.2*):

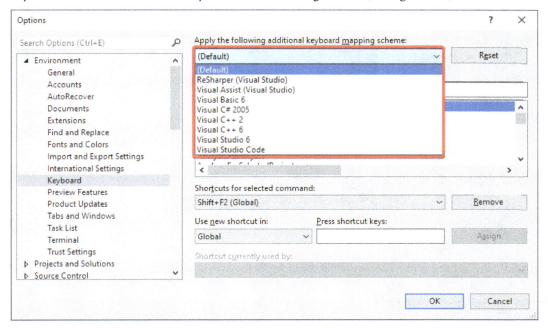

Figure 13.2 – Mapping schemes for shortcuts

To create a new shortcut, select **(Default)** under **Apply the following additional keyboard mapping scheme**, and then select the **Analyze.RunDefaultCodeCleanUpOnSolution** command from the list. This command executes a process that cleans up code to improve the format and removes unnecessary code. Finally, you can assign a custom shortcut for this command by pressing a key combination in the **Press shortcut keys** box – for example, the *Alt + C* combination (see *Figure 13.3*):

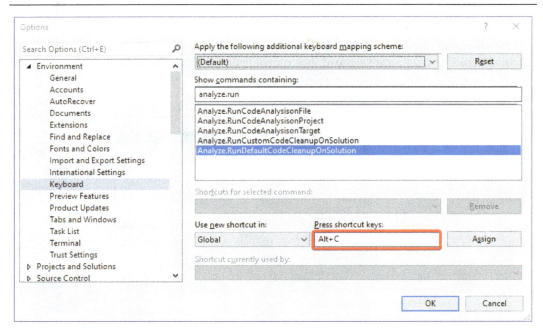

Figure 13.3 – Assigning the Ctrl + Alt shortcut to a command

Now, click **Assign** and then **OK** to confirm and add this new shortcut in VS (see *Figure 13.4*):

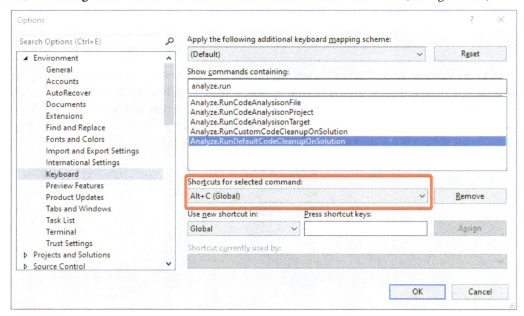

Figure 13.4 – The shortcut has been assigned to the command

After assigning the shortcut for the selected command, you can try it in VS. In this case, the **Analyze. RunDefaultCodeCleanUpOnSolution** command can be used globally, so you can have any project open, and by pressing *Ctrl + C*, VS will perform a cleanup on the solution.

> **Important note**
> You can override shortcuts included by default in VS. However, this is not the best practice since you are altering the normal behavior in VS, and it could be difficult to work in other environments.

In this chapter, we learned about shortcuts in VS, and we reviewed all the useful shortcuts that we can use while coding or performing an action with tools or functionalities.

With this knowledge of how to use shortcuts in your daily work, you will be able to depend less and less on your keyboard to execute actions within VS, which will allow you to become an efficient developer.

We also learned how to create shortcuts and automate common tasks in our projects using a key combination.

Farewell

It's been a long and exciting road from where we started. In the first section of this book, you learned how to install and adjust VS to suit your needs. We also explored the main templates for different types of projects that you can work with in VS. Similarly, we devoted an entire chapter to analyzing the various available debugging and profiling tools.

In the second section, you learned how to use Code Snippets to quickly create repetitive pieces of code, how to use GitHub Copilot as your programming companion, how to exploit tools for frontend and backend development, how to perform code cleanup, and studied different ways to publish projects for multiple platforms in VS 2022.

Finally, in the third section, you learned about how to integrate VS with GitHub, how to conduct live sessions with your project partners using Live Share, how to work with extensions, the most popular extensions and their uses, and the most important shortcuts that can help you perform repetitive tasks within VS.

At this point, we must congratulate you for reaching the end of this book. We hope you enjoyed reading it as much as we enjoyed writing it. The next step is to apply the knowledge you've acquired in your day-to-day work as a developer.

Happy coding!

Index

W

packtpub.com

Subscribe to our online digital library for full access to over 7,000 books and videos, as well as industry leading tools to help you plan your personal development and advance your career. For more information, please visit our website.

Why subscribe?

- Spend less time learning and more time coding with practical eBooks and Videos from over 4,000 industry professionals

- Improve your learning with Skill Plans built especially for you

- Get a free eBook or video every month

- Fully searchable for easy access to vital information

- Copy and paste, print, and bookmark content

Did you know that Packt offers eBook versions of every book published, with PDF and ePub files available? You can upgrade to the eBook version at packtpub.com and as a print book customer, you are entitled to a discount on the eBook copy. Get in touch with us at customercare@packtpub.com for more details.

At www.packtpub.com, you can also read a collection of free technical articles, sign up for a range of free newsletters, and receive exclusive discounts and offers on Packt books and eBooks.

Other Books You May Enjoy

If you enjoyed this book, you may be interested in these other books by Packt:

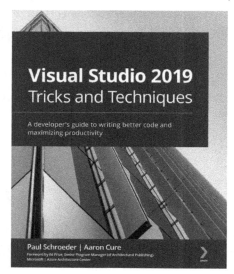

Visual Studio 2019 Tricks and Techniques

Jason Alls

ISBN: 978-1-80020-352-5

- Understand the similarities and differences between VS 2019 and VS Code
- Get to grips with numerous keyboard shortcuts to improve efficiency
- Discover IDE tips and tricks that make it easier to write code
- Experiment with code snippets that make it easier to write repeating code patterns
- Find out how to customize project and item templates with the help of hands-on exercises
- Use Visual Studio extensions for ease and improved productivity
- Delve into Visual Studio's behind the scene operations

High-Performance Programming in C# and .NET

Jason Alls

ISBN: 978-1-80056-471-8

- Use correct types and collections to enhance application performance
- Profile, benchmark, and identify performance issues with the codebase
- Explore how to best perform queries on LINQ to improve an application's performance
- Effectively utilize a number of CPUs and cores through asynchronous programming
- Build responsive user interfaces with WinForms, WPF, MAUI, and WinUI
- Benchmark ADO.NET, Entity Framework Core, and Dapper for data access
- Implement CQRS and event sourcing and build and deploy microservices

Packt is searching for authors like you

If you're interested in becoming an author for Packt, please visit authors.packtpub.com and apply today. We have worked with thousands of developers and tech professionals, just like you, to help them share their insight with the global tech community. You can make a general application, apply for a specific hot topic that we are recruiting an author for, or submit your own idea.

Hi!

We're Hector Uriel Perez Rojas and Miguel Angel Teheran Garcia, the authors of Hands-On Visual Studio 2022. We really hope you enjoyed reading this book and found it useful for increasing your productivity and efficiency in Visual Studio 2022.

It would really help us (and other potential readers!) if you could leave a review on Amazon sharing your thoughts on Hands-On Visual Studio 2022 here.

Go to the link below or scan the QR code to leave your review:

```
https://packt.link/r/1835080448
```

Your review will help us to understand what's worked well in this book, and what could be improved upon for future editions, so it really is appreciated.

Best Wishes,

Hector Uriel Perez Rojas

Miguel Angel Teheran Garcia

Download a free PDF copy of this book

Thanks for purchasing this book!

Do you like to read on the go but are unable to carry your print books everywhere?

Is your eBook purchase not compatible with the device of your choice?

Don't worry, now with every Packt book you get a DRM-free PDF version of that book at no cost.

Read anywhere, any place, on any device. Search, copy, and paste code from your favorite technical books directly into your application.

The perks don't stop there, you can get exclusive access to discounts, newsletters, and great free content in your inbox daily

Follow these simple steps to get the benefits:

1. Scan the QR code or visit the link below

https://packt.link/free-ebook/9781835080443

2. Submit your proof of purchase

3. That's it! We'll send your free PDF and other benefits to your email directly

www.ingramcontent.com/pod-product-compliance
Lightning Source LLC
Chambersburg PA
CBHW080621060326
40690CB00021B/4773